ADVANCED LEVEL

Data and Data Handling for A2 Level

BIOLOGY

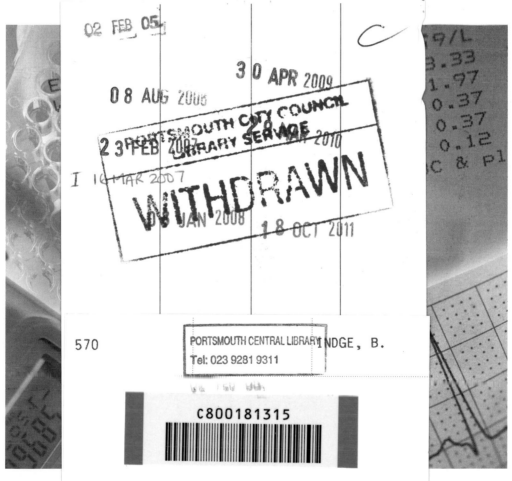

Portsmouth
CITY COUNCIL
LEISURE SERVICE

CL-1

Hodder & Stoughton

A MEMBER OF THE HODDER HEADLINE GROUP

D0244045

Orders: please contact Bookpoint Ltd, 130 Milton Park, Abingdon, Oxon OX14 4SB. Telephone: (44) 01235 827720. Fax: (44) 01235 400454. Lines are open from 9.00 – 6.00, Monday to Saturday, with a 24 hour message answering service. You can also order through our website www.hodderheadline.co.uk.

British Library Cataloguing in Publication Data
A catalogue record for this title is available from the British Library

ISBN 0 340 812753

First Published 2004
Impression number 10 9 8 7 6 5 4 3 2 1
Year 2010 2009 2008 2007 2006 2005 2004

Papers used in this book are natural, renewable and recyclable products. They are made from wood grown in sustainable forests. The logging and manufacturing processes conform to the environmental regulations of the country of origin.

Illustrations by Tech-Set and Peter Bull Art Studio.
Typeset in 11/16pt Sabon by Tech-Set Ltd, Gateshead, Tyne & Wear.
Printed in Great Britain for Hodder & Stoughton Educational, a division of Hodder Headline, 338 Euston Road, London NW1 3BH by Martins the Printers, Berwick upon Tweed.

Contents

Introduction

Biology is a science and, as such, it involves a lot more than learning a set of facts. A biologist needs a range of skills with which to investigate problems. The purpose of this book is to help you to master these skills, particularly those relating to the analysis of data. It builds on the foundations which were laid in **Data and Data Handling for AS Level Biology**. If you look through the pages of this book, however, you should see some differences in the style of the questions and the nature of the particular tasks you are required to carry out.

- The data with which you are provided at A2 are generally more complex. At AS you considered situations where there were just two variables – an independent variable and a dependent variable. Much of the information in this book involves situations where there are three or more variables so tables have more columns and graphs have more curves. This should not produce major difficulties but it will mean that you have to spend a little more time understanding the data before you attempt the questions that follow.

- The questions themselves are also somewhat more demanding. You will still encounter some which will require you to identify and describe patterns and trends. You will find others where you will need to draw on your knowledge of biology to explain and interpret aspects of the data. But you will also encounter more questions which require evaluation. Evaluation means 'judging the worth of' and you will be required to look critically at the techniques investigators used and the results they obtained, and judge their worth.

- From a teaching or learning point of view, it is often better to divide biology into separate topics and deal with each one independently. The more you study the subject, however, the more you will come to appreciate that this division is rather artificial. There are threads which link these topics together. Consider, for example, the processes by which substances pass into and out of cells . . . diffusion, osmosis and active transport. These may be basic AS topics but they underpin much of the physiology that is studied in A2 courses. You will meet these topics when you look at how the products of digestion are absorbed from the gut, how ions pass through membranes when a nerve impulse is initiated and transmitted along the axon of a nerve cell, and how urine is formed in the kidney. Because of this, the exercises in this book have not been confined to single topics – a knowledge of the basic principles that formed part of your AS course is often required as well.

• The last difference which you will encounter concerns the introduction of statistical tests. The results of any practical investigation which you undertake, whether in a school or college or in a university research laboratory, can be explained in one of two ways. There may be an underlying biological cause, or the results may simply be a matter of chance. As you cannot be sure, you need to carry out a suitable statistical test. This will enable you to assess the probability of a particular set of results being due to chance. None of the exercises in this book require you to carry out specific statistical tests but they do include questions which will check your understanding of the principles involved.

About this book

This book starts with a chapter that is designed to build on the basic experimental and analytical skills that formed an important part of your AS course. It introduces you to the idea of statistical tests and explains how they are used to analyse data and the important role they play in planning investigations.

Chapters 2 – 5 follow. They have a common format. There is usually an introduction. This points out features of the topic concerned which either make it particularly challenging or should be considered when answering questions on this subject. The exercises which follow illustrate themes which are common to all A Level Biology and Human Biology specifications, but a word of caution here. Although all specifications which have Biology in their title are based on the same core material, there is some variation in actual subject content so it is possible that some of these exercises will cover aspects of the topic concerned which are not part of your specification.

The material chosen as a basis for these exercises covers a range of topics from ions to ecosystems, and earthworms to kangaroos. Hopefully, this will make the exercises interesting. It should also provide you with opportunities to practise another important skill. Look at the section at the front of your specification which is headed 'Assessment Objectives' and particularly at Assessment Objective 2. This not only requires you to be able to interpret data in tables and graphs but also to 'apply biological principles and concepts in solving problems in unfamiliar situations'.

Using material that you have not encountered before will help you to develop the skills necessary to meet this important assessment objective.

Modular specifications have some real advantages over those which only examine candidates at the end of a course. They allow you to divide the subject into convenient and manageable chunks of knowledge. However, there is a disadvantage in this approach as it makes it more difficult to appreciate how different topics link to each other. Because of this all A Level specifications require the examiners to set questions which bring together ideas from different modules. These are sometimes referred to as synoptic questions. The exercises in Chapter 6 involve synoptic questions. They should be attempted as you near the end of your A2 course. Each of them explores a range of topics and requires you to bring these topics together and apply them in a particular context – another skill critical for A Level success.

Before you start

The questions in the exercises in this book are very similar to A Level questions and they require the same approach. Before you start, you must analyse the question and make sure that you understand what you are required to do. There is little point in spending a lot of time on these exercises if you don't follow the instructions given in the individual questions. Analysing questions requires you to do three things.

1 **Look at the command word**. This a word such as 'describe' or 'calculate' or 'explain' which tells you what you are required to do. The Box in this introduction contains a list of the important command words that have been used in writing the questions in this book. Use it from the start to make sure that you get into the habit of giving the required answer. The one expression that you will not find in this box is 'Write all you know about . . .' We will not use it here, and nor will your examiners!

2 **Make sure you know what you are required to write about**. The rest of the question will tell you this.

3 **Note the mark allocation**. Mark allocations can be very helpful. In general, in data handling questions, one mark means you are required to do one thing or recall one piece of information that you have learnt during your A Level course. So, get into good habits now. Before you answer any question ask yourself

- What have I got to do?
- What have I got to write about?
- How many marks are available?

It does not take very long, but it will help you to turn your knowledge and understanding into examination marks when the time comes.

BOX 1 Instructions used in data-handling questions

- **Calculate** means calculate! Although no explanation ought to be required for this term, an important point should be made. Make sure that you show your method of working as clearly as possible. In some cases it is possible to get marks for the right approach, even if the actual answer is wrong. You will only get marks, however, if you explain your method clearly enough.

- **Describe** involves giving a written description of information presented in a table or graph. All that you need to do is to summarize the main trends or patterns and relate these to the figures provided. Answers such as 'The rate of reaction increases steadily to a peak value of 4.2 cm^3 h^{-1} at 50°C. It then falls to zero at 65°C' should gain full credit. An explanation is not required.

- **Explain** means give a reason why. A description is not required and will not gain credit. Perhaps the best check is to ask yourself, 'Have I explained why ... ?' In A Level examinations more marks are lost through failing to give proper explanations than for almost any other reason.

- **Give evidence from/Using examples from** involves making use of the material provided to illustrate a particular point. Since this is a requirement of the question, full marks will not be given for a general answer which fails to refer to the relevant material.

- **Sketch** is used where a curve has to be added to a graph. When this term is used, it is simply the shape of the curve that is required. There is no need to invent figures and attempt to plot these accurately.

- **Suggest** is used where it is expected that you will not be able to answer from memory. There may be more than one valid answer and, in general, any sensible response based on sound biological reasoning will be acceptable.

Using statistical tests

Why do we use statistical tests?

Look at a roadside verge in early spring and you will almost certainly see dandelions in flower. Examine a particular length of verge more closely and you will probably see that most of these dandelions are near the road. Farther away, they are less common and other species of plant are more numerous. How can we explain this observation? There are a number of possible biological explanations. Perhaps the road was treated with salt during the winter and dandelions can tolerate high salt concentrations better than other plants. Perhaps they can tolerate frequent mowing or trampling. There is another possible explanation, however, which has very little to do with biology. Maybe it is all due to **chance**. It could be that we just happened to pick an area where there were more dandelions growing closer to the road.

We will consider a particular investigation in more detail. Observations suggest that birds are able to distinguish between different colours. Insect-eating birds learn to avoid distasteful species such as ladybirds. Many ladybirds are bright red. The berries of many plants turn red as they ripen. Blackbirds eat ripe, red hawthorn berries but do not eat unripe, green ones. Slug pellets, which gardeners use to kill slugs and snails, are coloured blue. The manufacturers claim that birds avoid blue so will not be poisoned by feeding on slug pellets.

Colour choice by blackbirds was investigated. Oatmeal was dyed one of four different colours: red, green, yellow or blue. The dyed oatmeal was dipped in fat to make it more attractive as bird food and put out on a tray. Records were kept of birds visiting the tray and the colours of the food they selected and ate. Each week the results of 50 visits were recorded. Some results from this investigation are shown in Table 1.1.

Table 1.1 Colour preference in blackbirds

| Week | Number of times blackbirds fed on food coloured | | | |
	Red	Green	Yellow	Blue
1	32	13	5	0
2	38	3	7	2
3	29	6	14	1
4	36	2	12	0
Mean	33.8	6.0	9.5	0.8

What conclusions can be drawn from these data? Assuming that the behaviour of blackbirds in this investigation is linked to their feeding behaviour with normal food, it would seem fair to suggest that there is a distinct preference for red and that they seldom select blue. We would have little hesitation in coming to this conclusion.

Now look at the difference between blackbirds feeding on yellow-coloured food and green-coloured food. It is not easy to arrive at a firm conclusion from the evidence in the table. Although the mean value for those eating yellow food is higher than for those eating green food, the difference is not large and, in addition, there is considerable variation from one week to the next. In fact, in the first week, more blackbirds fed on green food than on yellow food. The difference between the results may simply be due to chance. We need a more precise way of looking at our results than just relying on personal opinion. We must apply an appropriate **statistical test**. This allows us to calculate the probability of obtaining results that differ from each other by a particular amount purely by chance.

In this particular case, carrying out a suitable statistical test allows us to make the statement that the probability of blackbirds selecting yellow rather than green food is approximately 19 out of 20. We could turn this statement round and say that there is only a probability of 1 in 20 that these results have arisen by chance.

We accept different levels of probability in different situations. Suppose you were offered a 10p raffle ticket in a draw for which the prize was a car. You would probably be very happy to buy a ticket if there was a 1 in 20 chance of winning the car! You might even be very tempted to buy a ticket if the probability of winning the car was 1 in 100 or even 1 in 1000. Now look at another situation. Suppose investigators established a strong link between eating parsnips and dying of cancer. You would probably be very reluctant to accept a risk of 1 in 1000 and parsnips would rapidly disappear from the menu!

As biologists, we understand that we can never completely rule out the effect of chance. Look back again at Table 1.1. We felt quite confident in saying that on this evidence blackbirds prefer red food to green food but there is still a faint probability that the results we obtained were due to chance – in fact, this probability is a lot less than 1 in 1000. It seems quite reasonable to say that the probability of results like these arising due to chance is so low that we can safely reject it as an explanation. We need a cut-off point and for biological experiments we normally accept this as a probability of 1 in 20 or 0.05. If there is a probability of less than 1 in 20 that our results could have arisen by chance then we recognise that there must be another explanation for them.

Different statistical tests for different purposes

A statistical test is a tool and, like all tools, you don't need to know how it works to make use of it. If you have an electric drill or a food processor, it is not necessary to understand how an electric motor works in order to use it. What you do need to be able to do, however, is to select the right tool for the right job. Statistical tests involve making the same decisions and the first of these is to decide which test you need to use.

The investigations which you carry out during your A Level Biology course generate data which you may need to look at in different ways. In general terms, you are most likely to want to do one of the following

- See if there is a difference between two sets of data. Suppose you were looking at the effect of pH on the rate of reaction of an enzyme. Are the data you obtained at a pH of 5 significantly different from those you obtained at a pH of 8?

- See if there is an association between two sets of data. This is often what you want to do in an ecological investigation. Is there a significant association between, say, the number of sea aster plants and the salinity of the soil?

- The data you have collected may fall into distinct categories. If this is the case you may want to know if the difference between the data you have collected and those which you have predicted is significant.

Each of these situations demands a different statistical test. The decision chart in Figure 1.1 should help you choose which one to use. There are many different statistical tests. This chart includes all the tests an A Level Biologist is likely to need. You should look carefully at this chart together with your specification. Because different specifications require the use of different tests, you can amend the chart to suit your purposes. It is unlikely that you will need to add anything, but you may well be able to simplify it.

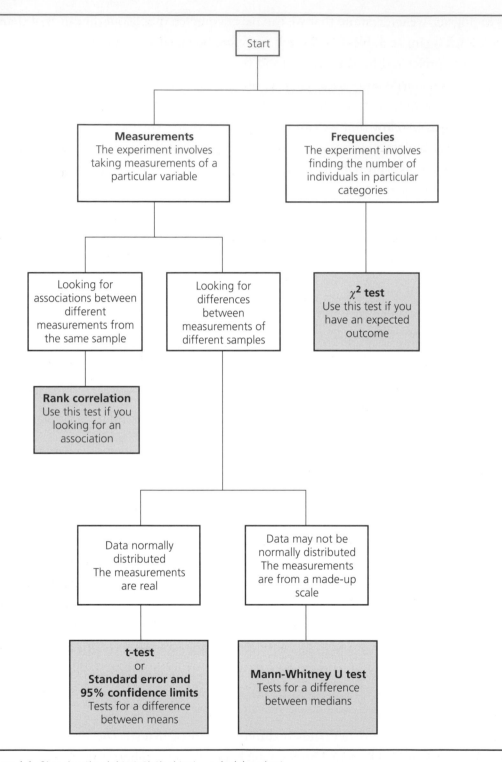

Figure 1.1 *Choosing the right statistical test – a decision chart*

Making this decision is extremely important. The statistical test you use and the way you design your investigation go hand in hand. If you want to put up a shelf, the decision to use either screws or nails will determine whether you need a screwdriver or a hammer. In the same way, it is essential when designing an investigation to bear in mind the statistical tests you might wish to use. In this way you can be sure that you collect sufficient data in a form which can be tested.

Carrying out statistical tests

The null hypothesis

All statistical tests that you might use at A Level centre round what we refer to as a **null hypothesis**. It is obviously important, therefore, to understand what is meant by a null hypothesis and why one is used. We will consider an actual example. Suppose you were investigating the effect of two different temperatures on the loss of pigment from beetroot tissue. You need to have a clear hypothesis that you can test. It is difficult to predict what will happen – after all, this is an investigation so you probably don't know. It is much easier to phrase our hypothesis in terms of what will happen if there is no difference or no association. This is called a null hypothesis. In the example above, the null hypothesis would be that there is no difference in the amount of pigment lost at these temperatures. You can then carry out your test and, as a result, come to a decision about whether or not to accept the null hypothesis. If the test indicates that you should accept it, then all you can say is that there is no significant difference between the results at the two temperatures. We also have to accept that these results could well be explained by chance. If we reject the null hypothesis, we can say that these results are unlikely to be due to chance and can accept that there is a biological explanation.

Calculating the test statistic

Having set up a null hypothesis you need to carry out the test you have selected. The best way to look at this is to use another analogy. Suppose you wanted to make a coffee cake from scratch. You would need a carbohydrate base, flavouring and something to bind the ingredients together but, because you probably have a somewhat limited understanding of the theory behind cake making, your cake is not likely to be a success. So, you decide to use a recipe book. You look under 'cakes' but, possibly, do not find exactly what you want and so decide to amend a chocolate cake recipe by substituting coffee for chocolate. You may well end up with an excellent cake even though you understood little of the underlying theory.

As a biologist, carrying out statistical tests is like using a recipe book when making a cake. Find the test which does what you want, then amend it by substituting your data for those in the example. If you work methodically, and can use your calculator, there is nothing to be apprehensive about. It is all fairly straightforward. The boxes which follow the next section can act as your recipe book. They provide worked examples of the more common statistical tests.

Interpreting the results

Calculating the test statistic gives you a value. Look in Box 2 which contains a worked example of the χ^2 test. It shows that, in this case, the value of χ^2 is 6.34. What does this mean? You need to interpret this figure in terms of probability. You can then make the decision whether to accept or reject your null hypothesis. In order to do this you need to look up the figure you have calculated in a set of statistical tables, where the probability values have already been calculated for you.

It is important to bear in mind, however, that the confidence you can place on your judgement will vary according to how many samples you have taken. You can be a lot more confident with a large sample than with a small one. **Degrees of freedom** take this into account. We also need to understand something about probability if we are to make sense of statistical tables. **Probability** is expressed quantitatively. The probability of an impossible event is zero. The probability of an event which is certain to happen is one. Mathematically, we express these values as p = 0 and p = 1. We made the point earlier in this chapter that, for biological investigations, if an event happens 19 out of 20 times we will accept this as support for our hypothesis. A value of 19 out of 20 gives us a probability of 0.95 (p = 0.95). This is our critical value.

So, we need to look up the results of our calculation in a table of critical values. We will work through the example shown in Box 2. The value of χ^2 that we calculated was 6.34. We now need to work out the number of degrees of freedom in order to look in the correct row in the table. In this case, the number of degrees of freedom is one less than the number of values of $\dfrac{(O - E)^2}{E}$ that we calculated. This is $3 - 1 = 2$. Now look in Table 1.2, which is an abbreviated version of the table in Box 2.

You can see from this table that our calculated value of χ^2 is larger than the critical value of 5.99 for 2 degrees of freedom. If the calculated value is equal to or greater

Table 1.2 Table showing critical values of χ^2 for different degrees of freedom

Number of degrees of freedom	Critical value
1	3.84
2	5.99
3	7.82
4	9.49
5	11.07

than this critical value then we reject the null hypothesis and say that there is a significant difference between the observed and expected values. If on the other hand, the critical value of χ^2 turned out to be less than the critical value, then we would have to accept our null hypothesis that there was no difference between the observed and expected values.

If you look in a statistics book you will see that most statistical tables are much more complex than those we have included in the boxes in this chapter. We have been able to simplify them because we are only really concerned with one value of p. Obviously, if you go on to study biology at a higher level you will need to use more complex tables but, at A Level, an approach based on the critical values given in the boxes is quite acceptable.

BOX 2 The χ^2 test

Use this test when

■ the measurements relate to the number of individuals in particular categories.

Three peanut feeders were set up

A was hung from a hook with elastic bands

B was hung from a hook with a loop of string

C was anchored securely to an upright post.

The different species of birds feeding at each feeder were recorded. The results for one species are shown in Table 1.3.

Table 1.3 Numbers of blue tits feeding on peanut feeders suspended in different ways

Species	Total number of birds observed feeding		
	Suspended by elastic bands	Suspended by string	Secured to fence post
Blue tit	53	43	30

The chi-square (χ^2) test is based on calculating the value of χ^2 from the equation

$$\chi^2 = \frac{\Sigma(O - E)^2}{E}$$

where O represents the observed results and E represents the results we expect.

Set up a null hypothesis

The first thing to do is produce a null hypothesis. With the χ^2 test, the null hypothesis is always 'There is no difference between the observed results and the expected results.'

Work out the values of the observed results and the expected results

The results in the table give us the observed results (O). We have to calculate the expected results (E). If the feeding behaviour of blue tits on peanut feeders was purely a matter of chance, we would expect equal numbers to visit each type of feeder. So the expected number visiting each type of feeder is

$$\frac{53 + 43 + 30}{3} = 42$$

Calculate χ^2

We now use the equation to calculate the value of χ^2. The best way to approach this is to write the values in a table such as Table 1.4 below.

Table 1.4 Calculating χ^2

	Type of feeder		
	Suspended by elastic bands	Suspended by string	Secured to fence post
Observed results (O)	53	43	30
Expected results (E)	42	42	42
$(O - E)^2$	121	1	144
$\dfrac{(O - E)^2}{E}$	2.89	0.02	3.43

In the equation, the symbol Σ means 'the sum of' so we have to add together all the separate values of $\dfrac{(O - E)^2}{E}$. In this example χ^2 is therefore 6.34

Look up and interpret the value of χ^2

Look at Table 1.5. This is a table of values of χ^2. Work out the number of degrees of freedom in order to look in the right row of the table. The number of degrees of freedom is one less than the number of values of $\dfrac{(O - E)^2}{E}$ that we calculated. In this case it is 2.

With 2 degrees of freedom, the value of χ^2 which we calculated is larger than 5.99.

The calculated value of χ^2 is greater than the critical value so we reject the null hypothesis and say that there is a significant difference between the observed and expected values.

Table 1.5 A table showing the critical values of χ^2 for different degrees of freedom

Degrees of freedom	Critical value
1	3.84
2	5.99
3	7.82
4	9.49
5	11.07
6	12.59
7	14.07
8	15.51
9	16.92
10	18.31

BOX 3 | Spearman's rank correlation test

Use this test when
- you wish to find out if there is a significant association between two sets of measurements
- you have between 7 and 30 pairs of measurements.

Great tits are small birds. In a study of growth in great tits the relationship between the mass of the eggs and the mass of the young bird on hatching was investigated. Some of the data collected are given in Table 1.6.

Table 1.6 The mass of great tit eggs and of the chicks on hatching

Mass of egg/g	1.37	1.49	1.56	1.70	1.72	1.79	1.93
Mass of chick on hatching/g	0.97	1.01	1.18	1.16	1.21	1.27	1.75

Set up a null hypothesis

The first thing to do is to produce a null hypothesis. 'There is no association between the mass of the eggs and the mass of the chicks which hatch from them.'

Calculate the correlation coefficient

Start by ranking the egg mass and the mass of the chicks (Table 1.7). Note that when two or more values are of equal rank, each of the values is given the average of the ranks which would otherwise have been allocated.
Calculate the difference between the rank values and square this difference.

Table 1.7 Rank values of egg mass and chick mass

Individual number	Egg mass/g	Rank	Chick mass/g	Rank	Difference in rank (D)	D²
1	1.37	1	0.99	1.5	0.5	0.25
2	1.49	2	0.99	1.5	0.5	0.25
3	1.56	3	1.18	5	2	4
4	1.70	4	1.16	3	1	1
5	1.72	5	1.17	4	1	1
6	1.79	6	1.27	6	0	0
7	1.93	7	1.75	7	0	0

Find the sum of the squares of the differences $\Sigma D^2 = 6.5$

Now calculate the value of the Spearman's rank correlation, R_S, from the equation

$$R_S = 1 - \frac{6 \times \Sigma D^2}{n^3 - n}$$

where n is the number of pairs of items in the sample

$$R_S = 1 - \frac{6 \times 6.5}{7^3 - 7}$$

$$= 1 - \frac{39}{336}$$

$$= 1 - 0.116$$

$$= 0.884$$

Look up and interpret the values of R_S

Note that the value of R_S will always be between 0 and either +1 or −1. A positive value indicates a positive association between the variables concerned. A negative value shows a negative association.

Table 1.8 is a table of values of R_S. Look in the table under the right number of pairs of measurements. With 7 pairs of values, the calculated value of R_S is larger than 0.79.

The calculated value of R_S is greater than the critical value so we reject the null hypothesis and say that there is a significant correlation between the mass of an egg and the mass of the chick that hatches from it.

Table 1.8 A table showing the critical values of R_S for different numbers of paired values

Number of pairs of measurements	Critical value
5	1.00
6	0.89
7	0.79
8	0.74
9	0.68
10	0.65
12	0.59
14	0.54
16	0.51
18	0.48

BOX 4 The Mann–Whitney U test

Use this test when

■ you wish to find out if there is a significant difference between two medians

■ the data are not normally distributed

■ the sample size is between 5 and 25 but the size of the samples might differ.

Table 1.9 shows the total number of rabbits seen along a transect on different occasions in two months of the year.

Table 1.9 The number of rabbits seen along a transect in May and in July

Occasion number	1	2	3	4	5	6
Number of rabbits seen in May	23	35	40	47	14	40
Number of rabbits seen in July	20	19	28	27	38	

Set up a null hypothesis

The first thing to do is to produce a null hypothesis. 'There is no difference between the number of rabbits seen along the transect in May and the number seen in July.'

Calculate the values of U_1 and U_2

These values are calculated from the formulae below:

$$U_1 = n_1 \times n_2 + \tfrac{1}{2}n_2(n_2 + 1) - \Sigma R_2$$
$$U_2 = n_1 \times n_2 + \tfrac{1}{2}n_1(n_1 + 1) - \Sigma R_1$$

Where n_1 = number of measurements in first sample

n_2 = number of measurements in second sample

ΣR_1 = total of rank values for first sample

ΣR_2 = total of rank values for second sample

In order to calculate R_1 and R_2 you need to combine the data into a single rank order.

Table 1.10 Combining data to calculate R_1 and R_2

												ΣR_1	ΣR_2
Rank	1			4			7		9.5	9.5	11	42	
May	14			23			35		40	40	47		
July		19	20		27	28		38					
Rank		2	3		5	6		8					24

Using the following values

n_1 = number of measurements in first sample   = 6

n_2 = number of measurements in second sample =   5

ΣR_1 = total of rank values for first sample   = 42

ΣR_2 = total of rank values for second sample   = 24

$U_1 = n_1 \times n_2 + \frac{1}{2}n_2(n_2 + 1) - \Sigma R_2$

$\quad = 6 \times 5 + \frac{1}{2}5(5 + 1) - 24$

$\quad = 30 + 15 - 24$

$\quad = 21$

and $U_2 = n_1 \times n_2 + \frac{1}{2}n_1(n_1 + 1) - \Sigma R_1$

$\quad\quad = 6 \times 5 + \frac{1}{2}6(6 + 1) - 42$

$\quad\quad = 30 + 21 - 42$

$\quad\quad = 9$

Look up and interpret the values of U

Compare the smallest value of U with the critical values of U shown in Table 1.11. Look in the table under the right number of pairs of measurements. With 5 pairs of values the smallest value of U is greater than the critical value of 2.

In this test, if the calculated value of U is greater than the critical value, we accept the null hypothesis and say that there is no significant difference between the number of rabbits seen along the transect in May and the number seen in July. We reject the null hypothesis if the calculated value of U is less than the critical value.

Table 1.11 A table showing the critical values of U for different degrees of freedom.

Number of pairs of measurements	Critical value
5	2
6	5
7	8
9	13
10	23
11	30
12	37
13	45
14	55
15	64

BOX 5 Using standard error and 95% confidence limits

Use this test when

■ you wish to find out if there is a significant difference between two means
■ the data are normally distributed
■ the sizes of the samples are at least 30.

Biologists investigated the effect of the distance apart that parsnip seeds were planted on the number of seeds that germinated. Sets of 20 parsnip seeds were grown in trays. The seeds in a set were either touching each other or placed 2 cm apart. The numbers of seeds in each set that had germinated after 10 days were recorded and are shown in the Table 1.12.

Table 1.12 The effect of distance apart on the germination of parsnip seeds

Number of seeds that had germinated after 10 days											
Seeds touching each other						Seeds placed 2 cm apart					
8	10	10	5	6	12	14	16	16	12	15	16
15	9	13	12	9	10	10	12	13	16	12	10
8	10	7	10	8	7	15	11	15	14	13	13
14	9	11	7	9	14	11	15	14	8	13	14
11	10	9	12	8	9	19	11	12	17	9	17

Set up a null hypothesis

The first thing to do is to produce a null hypothesis. 'There is no difference between the number of seeds which germinated when they were touching than when they were placed 2 cm apart'.

Calculating standard error

Use a calculator to work out the mean and standard deviation of each of the samples. These values are shown in Table 1.13.

Table 1.13 Mean and standard deviation for germinating seeds

	Seeds touching each other	Seeds placed 2 cm apart
Mean	9.70	13.43
Standard deviation	5.60	6.45

Calculate the standard error of the mean, SE, for each sample from the following formula

$$SE = \frac{SD}{\sqrt{n-1}}$$

where SD = the standard deviation

and n = sample size

Table 1.14 shows all these values.

Table 1.14 Mean, standard deviation and standard error for germinating seeds

	Seeds touching each other	Seeds placed 2 cm apart
Mean	9.70	13.43
Standard deviation	5.60	6.45
Standard error	1.03	1.20

Interpreting the values

The 95% confidence limits are 1.96 standard errors above the mean and 1.96 standard errors below the mean (1.96 can be rounded up to 2.0). If the 95% confidence limits do not overlap, there is a 95% chance that the two means are different. In other words, there is a significant difference between the means at the 5% level of probability. We will now plot this information as a simple graph (see Figure 1.2). We have plotted the mean values for each of our samples as crosses and drawn bars to represent 2 standard errors on either side of each of these mean values. In this case, there is an overlap between the bars so we accept the null hypothesis and say that there is no significant difference between the number of seeds that germinated when they were touching and when they were placed 2 cm apart.

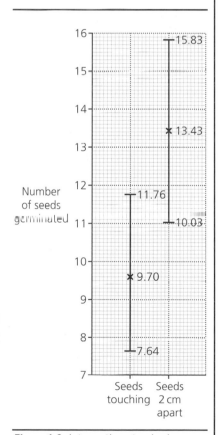

Figure 1.2 Interpreting standard error

BOX 6 Using the t test

Use this test when

- you wish to find out if there is a significant difference between two means
- the data are normally distributed
- the sample size is less than 25.

Biologists investigated the effect of the distance apart that parsnip seeds were planted on the number of seeds that germinated. Sets of 20 parsnip seeds were grown in trays. The seeds in a set were either touching each other or placed 2 cm apart. The numbers of seeds in each set that had germinated after 10 days were recorded and are shown in the Table 1.15.

Table 1.15 The effect of distance apart on the germination of parsnip seeds

Number of seeds that had germinated after 10 days											
Seeds touching each other						Seeds placed 2 cm apart					
8	10	10	5	6	12	14	16	16	12	15	16
15	9	13	12	9	10	10	12	13	16	12	10
8	10	7	10	8	7	15	11	15	14	13	13

Set up a null hypothesis

The first thing to do is to produce a null hypothesis. 'There is no difference between the number of seeds which germinated when they were touching than when they were placed 2 cm apart.'

Calculating the value of t

t can be calculated from the formula

$$t = \frac{\bar{x}_1 - \bar{x}_2}{\sqrt{(s_1^2/n_1) + (s_2^2/n_2)}}$$

where $\bar{x}_1$ = mean of first sample

$\bar{x}_2$ = mean of second sample

s_1 = standard deviation of first sample

s_2 = standard deviation of second sample

n_1 = number of measurements in first sample

n_2 = number of measurements in second sample

Use a calculator to work out the mean and standard deviation of each of the samples. These values are shown in Table 1.16.

Table 1.16 Mean, standard deviation and number of measurements for germinating seeds

	Seeds touching each other (1)	Seeds placed 2 cm apart (2)
Mean	9.15	12.79
Standard deviation	3.38	3.68
Standard error	18	18

Substituting these figures into the equation

$$t = 9.15 - \frac{12.79}{\sqrt{(11.42/18) + (13.54/18)}}$$

$$= \frac{-3.64}{\sqrt{0.63 + 0.75}}$$

$$= \frac{-3.64}{1.09}$$

$$= -3.34$$

Look up and interpret the values of t

Note that the value of t may be positive or negative. The sign does not matter so we can ignore this.

Look at Table 1.17. This is a table of values of t. Work out the number of degrees of freedom in order to look in the right row of the table. This is calculated using the formula $(n_1 + n_2) - 2$. In this case it is 34. Although the table does not give us an exact value, we can see that with 34 degrees of freedom, the value of t which we calculated would be larger than the critical value. This means we can reject the null hypothesis and say that there is a significant difference between the number of seeds that germinated when they were touching and when they were placed 2 cm apart.

Table 1.17 A table showing the critical values of t for different degrees of freedom.

Degrees of freedom	Critical value	Degrees of freedom	Critical value
5	2.57	15	2.13
6	2.48	16	2.12
7	2.37	18	2.10
8	2.31	20	2.09
9	2.26	22	2.07
10	2.23	24	2.06
11	2.20	26	2.06
12	2.18	28	2.05
13	2.16	30	2.04
14	2.15	40	2.02

Exercise 1.1 Which test?

1   Some investigations are described briefly below. For each of them:

   (a)   write a suitable title for the investigation

(1 mark)

   (b)   draw a table in which the results could be written

(2 marks)

   (c)   explain which of the statistical tests you would use.

(2 marks)

- In an ecological investigation, a series of quadrats, each measuring 25 cm × 25 cm were placed at random in an area of woodland. The bluebell plants growing in each quadrat were counted. Light meter readings from each quadrat were also taken.

- The behaviour of woodlice was investigated. The lids of two Petri dishes were joined together so that woodlice could go through a small gap from one to the other. Wet cotton wool was placed under a piece of gauze in the base of one dish. This produced humid conditions. Silica gel was placed under a piece of gauze in the base of the other dish. This produced dry conditions. Ten woodlice were placed in the apparatus and the distribution of the woodlice in the humid and dry sides was recorded over a period of five minutes.

- Daphnia is a small aquatic animal. When viewed with a microscope, its heart can be clearly seen and the number of beats per minute can be recorded. The water temperature in which Daphnia lives can be changed. In an investigation the heart rates of a large number of Daphnia were measured at three different temperatures, 15 °C, 19 °C and 23 °C.

- Hoverflies are important pollinators of flowers. In a rose bed, they appeared to visit yellow roses more frequently than roses of other colours. An experiment was set up to investigate this further. Pieces of yellow, red, green and blue card were laid out in a rose bed and the numbers of hoverflies approaching each piece in a 30 minute period were counted.

CHAPTER
two
Energy and energy transfer

Exercise 2.1 Respiration and mitochondria

Detailed studies of the biochemistry of respiration rely on isolating mitochondria from cells. Today, we can carry out this procedure in a school or college laboratory and it is easy to forget that the problems facing early workers were immense. In 1948, Kennedy and Lehninger carried out a series of experiments which helped to establish where, in a cell, the various processes associated with respiration took place. They worked with liver cells. Table 2.1 shows some of their results.

Table 2.1 Some results of the experiments carried out by Kennedy and Lehninger on the location of processes associated with respiration

Cell fraction	Rate of oxygen uptake/mmol mg^{-1} h^{-1} when following substrates added			Aldolase activity as percentage of activity in whole homogenate
	None	Citrate	Pyruvate	
Whole homogenate	0.5	n/a	6.8	100
Nuclei	0.0	1.9	0.9	3
Mitochondria	0.2	7.1	7.7	1
Ribosomes	0.2	0.5	0.4	n/a
Supernatant	n/a	n/a	n/a	96

n/a no results available

1 When these experiments were carried out, the techniques used today for obtaining pure cell fractions had not been developed. Give one piece of evidence from the table that shows the cell fractions were not pure. Explain your answer.

(2 marks)

2   Aldolase is an enzyme involved in respiration.

   (a)   What do the results in Table 2.1 show about the location of
          aldolase activity in a liver cell?

   (1 mark)

   (b)   What does your answer to part (a) suggest about the role of
          aldolase in respiration?

   (1 mark)

3   (a)   Suggest why oxygen was taken up by the mitochondria when no
          respiratory substrate was added.

   (1 mark)

   (b)   Citrate is formed when acetylcoenzyme A combines with a four-
          carbon compound at the start of the Krebs cycle. Use your
          knowledge of respiration to explain the rate of oxygen uptake by
          the mitochondria when citrate is added.

   (3 marks)

4   Is pyruvate able to pass through the outer membrane into a mitochondrion?
    Give evidence from the table to support your answer.

   (1 mark)

An oxygen electrode measures the oxygen concentration of a solution. A fresh
preparation of mitochondria was made and mixed with a suitable substrate. At
intervals, ADP was added. Figure 2.1 shows the trace obtained when this mixture
was monitored with an oxygen electrode.

Exercise 2.1 *continued*

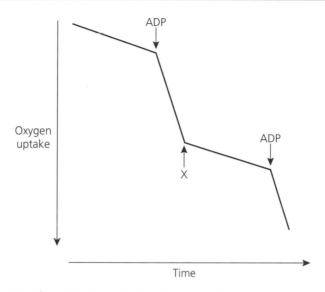

Figure 2.1 *Oxygen uptake by a mitochondrial preparation*

5   Suggest why glucose would not have been a suitable respiratory substrate in this investigation.

(1 mark)

6   (a)   Explain why the rate of oxygen uptake decreases at time **X**.

(1 mark)

(b)   What evidence is there from the trace that supports your answer to part (a)

(1 mark)

7   In a respiring cell, electron transfer is normally coupled with ATP synthesis. Without ATP synthesis, electron transfer will not take place. Dinitrophenol is a substance which uncouples these two processes. What would happen to the trace if dinitrophenol were added at time **X**? Give an explanation for your answer.

(3 marks)

Exercise 2.2 Respiratory Quotients

The Respiratory Quotient (RQ) is defined as the volume of carbon dioxide produced during respiration divided by the volume of oxygen consumed. In other words, it may be calculated from the formula

$$RQ = \frac{\text{Volume of carbon dioxide produced}}{\text{Volume of oxygen consumed}}$$

1   Carbohydrates have an RQ of 1. In an investigation, a woodlouse produced 1.4 cm^3 of carbon dioxide over a period of five minutes. What was its rate of oxygen consumption in cm^3 minute^{-1} if its respiratory substrate was carbohydrate?

(1 mark)

As the volume of a gas is proportional to the number of molecules present, the RQ can also be calculated from the relevant chemical equation. The equation below shows how the substance tripalmitin is oxidised during the process of respiration.

$$2C_3H_5(OOC.C_{15}H_{31})_3 + 145O_2 \longrightarrow 102CO_2 + 98H_2O$$

2   Tripalmitin is a triglyceride formed from glycerol and palmitic acid.

(a)   Use the equation to give the formula of palmitic acid.

(2 marks)

(b)   Is palmitic acid a saturated or an unsaturated fatty acid? Give an explanation for your answer.

(1 mark)

3   Calculate the RQ for tripalmitin. Show your working.

(2 marks)

Table 2.2 shows the RQ for different respiratory substrates.

Table 2.2 Respiratory Quotients for different respiratory substrates

Respiratory substrate	RQ
Carbohydrate	1.0
Protein	0.8 – 0.9
Triglyceride	0.7

Exercise 2.2 *continued*

4   Table 2.2 shows a range of values for the RQ of protein. Suggest an
    explanation for this.   *(2 marks)*

5   (a)   Vampire bats feed entirely on blood. Use information from
          Table 2.2 to suggest the likely RQ of a vampire bat which has
          recently fed. Give the reason for your answer.   *(2 marks)*

    (b)   During strenuous exercise some of the muscles in the body respire
          anaerobically. As a result lactate accumulates. This lactate causes
          some of the carbon dioxide present in the blood as
          hydrogencarbonate to be released. How would a period of
          strenuous exercise affect the RQ of a person respiring
          carbohydrate? Explain your answer.   *(3 marks)*

Figure 2.2 shows the respiratory substrates used by a locust during a prolonged
period of flight.

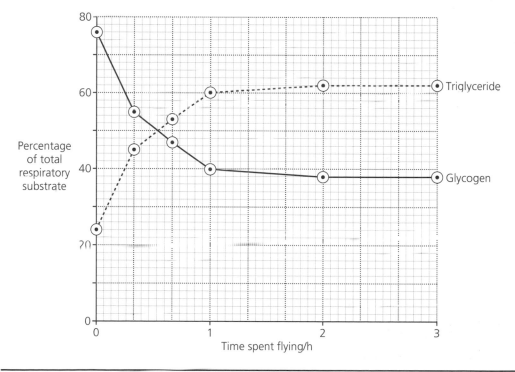

Figure 2.2 *The effect of a prolonged period of flight on the respiratory substrates used by a locust*

6   Describe how the RQ of the locust would change over the period of time
    shown on the graph.   *(2 marks)*

Exercise 2.3 Chloroplasts and photosynthesis.

Photosynthesis involves two groups of reactions, the light-dependent reactions, which generate ATP and reduced NADP, and the light-independent reactions in which carbon dioxide is converted to carbohydrate. These processes are summarised in Figure 2.3.

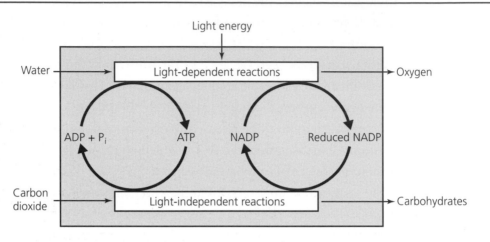

Figure 2.3 *A summary of the light-dependent and light-independent reactions of photosynthesis*

The existence of these separate groups of reactions had been suggested as long ago as 1905 by Blackman, a British plant physiologist. Blackman carried out a series of investigations as a result of which he identified a number of factors which limited the rate of photosynthesis. Figure 2.4 shows some results from one of these investigations.

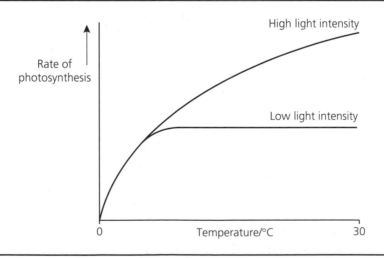

Figure 2.4 *The effect of light intensity and temperature on the rate of photosynthesis*

1   From his work, Blackman concluded that at low light intensities, light intensity limited the rate of photosynthesis; at high light intensities the rate was limited by temperature. Explain the evidence from Figure 2.4 that supports these conclusions.

(2 marks)

2   Blackman explained his results by suggesting that there was a light-dependent reaction unaffected by temperature, and a light-independent reaction which was temperature dependent. Use your knowledge of photosynthesis to explain why the light-independent reactions are temperature dependent.

(2 marks)

It was not until 50 years later that Arnon and his colleagues were able to show precisely where in a chloroplast these reactions took place. This work involved preparing a pure suspension of chloroplasts from plant cells by cell fractionation and centrifugation. Some of these chloroplasts were broken up and used to produce a suspension of chloroplast membranes. Table 2.3 shows the results of some investigations carried out on these suspensions.

Table 2.3 Some results from Arnon's investigations on chloroplasts

	Intact chloroplasts		Chloroplast membranes	
	Light	Dark	Light	Dark
Oxygen produced/ mm^3 $hour^{-1}$	480	0	400	0
Carbon dioxide taken up/arbitrary units	52000	238	40	37
Inorganic phosphate taken up/arbitrary units	4250	950	3100	300

3   (a)   Describe how you could use centrifugation to obtain a preparation of chloroplasts which is not contaminated by nuclei.

(2 marks)

    (b)   Explain why it is important that these chloroplast preparations are not contaminated with mitochondria.

(2 marks)

Exercise 2.3 *continued*

4   (a)   Give **two** pieces of evidence from Table 2.3 that show the light-dependent reactions take place on the chloroplast membranes.

(2 marks)

     (b)   Explain the evidence from Table 2.3 that the light-independent reactions take place in the stroma of the chloroplasts.

(2 marks)

5   Suggest why a small amount of carbon dioxide is taken up by the intact chloroplasts in the dark.   *(2 marks)*

Figure 2.5 is a flowchart summarising a further investigation which confirmed the site of the light-independent reactions.

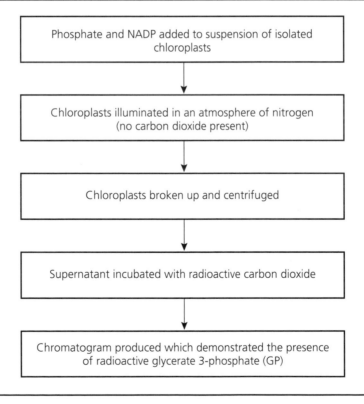

Figure 2.5

6   Name **one** substance other than phosphate and NADP which would need to be added to the chloroplast suspension to ensure that the light-dependent reactions take place.

(1 mark)

Exercise 2.3 *continued*

7 (a) What happened to the NADP added to the chloroplast suspension when it was illuminated?

(1 mark)

(b) Explain why it was necessary to exclude carbon dioxide when the chloroplasts were illuminated.

(2 marks)

8 Explain how the results of this investigation show that the light-independent reactions take place in the stroma of the chloroplast.

(2 marks)

Exercise 2.4 Producers

All crop plants rely on the process of photosynthesis to convert light energy to chemical potential energy. Consider a crop of grass growing in a field. A lot of light energy falls on this crop during the year. Only a small part of it, however, can be used by the grass and converted into chemical potential energy. Figure 2.6 shows what happens to the light energy falling on 1 m² of the crop. The units are $kJ\ m^{-2}\ year^{-1}$.

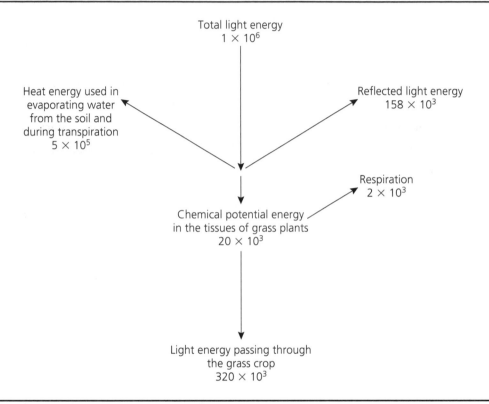

Figure 2.6 *The transfer of energy in a grass crop*

1   (a)   What is the maximum amount of chemical potential energy which could be passed from the grass to organisms at the next trophic level?

(1 mark)

(b)   Calculate the percentage of the total light energy transferred to chemical potential energy by photosynthesis. Show your working.

(3 marks)

Exercise 2.4 *continued*

2   Explain how each of the following may affect the percentage of the total light energy which is transferred to chemical potential energy in grass.

    (a)   Sowing the grass seed at a higher density so that the grass plants are closer together.

(2 marks)

    (b)   Growing grass on soil where the nitrate concentration is very low.

(2 marks)

The percentage of light energy which is converted to chemical potential energy during photosynthesis is low. Sugar cane is a crop grown in large plantations in hot, wet tropical regions. It is able to convert about 7% of the light energy it receives into chemical potential energy and is one of the most efficient of all plants in this respect.

3   Explain **one** way in which each of the following contribute to sugar cane being so efficient.

    (a)   It is grown in large plantations by commercial growers.

(2 marks)

    (b)   It grows in the tropics.

(2 marks)

    (c)   It has tall upright leaves rather than leaves arranged horizontally.

(2 marks)

Table 2.4 compares the productivity and the standing crop biomass of the producers in different ecosystems. Productivity is the amount of biomass produced per year. Standing crop biomass is the biomass present at the time the measurements were made.

Exercise 2.4 *continued*

Table 2.4 Productivity and standing crop biomass of the producers in different ecosystems

Ecosystem	Mean annual productivity/ g m^{-2}	Mean standing crop biomass/ kg m^{-2}
Tropical rain forest	2200	45
Deciduous woodland (UK)	1200	30
Evergreen woodland (UK)	1300	30
Cultivated land	650	1

4   Most of the crops grown on the cultivated land are annual plants, which are harvested at the end of the growing season. Explain how the figures in the table support this.

(2 marks)

5   (a)   Suggest an explanation for the difference between standing crop biomass in tropical rain forest and on cultivated land.

(1 mark)

(b)   The productivity of the producers in tropical rain forest is higher than that of producers on cultivated land in the tropics. Explain why.

(2 marks)

6   Suggest an explanation for the difference in mean productivity between deciduous and evergreen woodland in the UK.

(1 mark)

Exercise 2.5 Feeding pigs

Scientists working with farm animals have investigated what happens to the energy in food once it has been eaten. In this exercise we shall look at energy transfer in pigs. A pig differs from farm animals such as cows and sheep because it has a simple stomach and a digestive system very like that of a human. Figure 2.7 summarises what happens to the energy in food once it has been eaten by a pig.

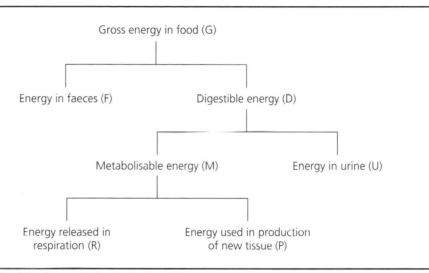

Figure 2.7 *The fate of energy in food eaten by a pig*

1   Pigs kept inside sheds during the winter months show a greater gain in mass than pigs kept outside when given the same amount of food. Use information from Figure 2.7 to explain why.

(2 marks)

2   Use appropriate letters from Figure 2.7 to write an equation showing

   (a)   how digestible energy is related to gross energy

(1 mark)

   (a)   how the energy used in production of new tissue is related to digestible energy.

(1 mark)

Exercise 2.4 *continued*

Table 2.5 shows energy values of some typical pig foods. Coconut cake meal is a by-product of the process in which edible oils are extracted from coconuts.

Table 2.5 Energy values for a range of pig foods

Food	Gross energy/ MJ kg^{-1}	Energy lost in faeces/MJ kg^{-1}	Energy lost in urine/MJ kg^{-1}	Metabolisable energy/MJ kg^{-1}
Maize	18.9	1.6	0.4	16.9
Oats	19.4	5.5	0.6	
Barley	17.5	2.8		14.2
Coconut cake meal	19.0	6.4	2.6	10.0

3   Calculate the two values missing in the table

(2 marks)

4   Coconut cake meal contains approximately 153 g kg^{-1} of crude fibre (cellulose and pectin) and 220 g kg^{-1} of protein. Maize contains 24 g kg^{-1} of crude fibre and 98 g kg^{-1} of protein. How would

   (a)   the difference in crude fibre

   (b)   the difference in protein

   explain why metabolisable energy in coconut cake meal is less than the metabolisable energy in maize?

(4 marks)

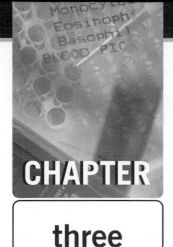

CHAPTER

three | **Ecology**

A chapter on ecological questions deserves a special introduction. A Level students often find that questions involving ecology present particular difficulties. In unit tests based on this area of biology, it is generally not difficult to scrape together enough marks to gain a pass. It is much more of a challenge to do well. Because of this, we will start this chapter by looking at three particular things you should keep in mind whenever you encounter questions based on ecological topics.

Language

In some ways, biology is like a foreign language. It has its own vocabulary with biological terms which have very specific meanings. Unfortunately, some of these terms are also in everyday use and this leads to confusion. Take, for example, the word 'organic'. To a biologist, an organic substance is one in which molecules are based on the element carbon. The word has been hijacked, however, and to a person buying vegetables, for example, in a supermarket it means something which has been produced without the use of chemical fertilisers or pesticides...a very different meaning.

The first rule of answering questions on ecological topics, whether or not they involve the handling of data, is to use the correct terms in the correct context. Animals and plants have distinct ecological niches: they do not have 'homes'. Weeds compete with crop plants for resources: they do not 'fight for room to grow'. Before you answer any of the questions in this chapter, look at Box 7. It summarises some of the basic ecological terms you should have encountered and should be using, whatever specification you are following.

BOX 7

A glossary of basic ecological terms. Make sure that you use these terms when answering ecological questions

Environment

- The environment of an organism is the set of conditions which surround it. The environment consists of a non-living or abiotic component and a living or biotic component.

Abiotic

- Abiotic factors are those which relate to the non-living part of the environment. Soil nutrient concentration, temperature and rainfall are all abiotic factors.

Biotic

- Biotic factors are those which relate to the living part of the environment. Competition and predation are examples of biotic factors.

Population

- A population is a group of individuals belonging to a particular species. The members of a particular population will be found in a particular place at the same time.

Community

- A community is the term used to describe all the populations of different organisms living in a particular place at a particular time.

Ecosystem

- An ecosystem is the basic unit of ecology. It consists of the community of living organisms and the abiotic factors which influence these organisms.

Habitat

- The habitat is the place where a particular population or community lives.

Niche

- The meaning of this term is rather complicated. In very simple terms, an organism's niche is the place where it is found and what it does there. It is a description of how an organism fits into its environment.

BOX 7 continued

Producer

■ A producer is an organism which can produce complex organic molecules from simple inorganic ones, using energy from sunlight or from chemical reactions. Most plants are producers.

Consumer

■ Consumers feed on other organisms. They get their nutrients from complex organic molecules. Primary consumers are herbivores and feed directly on producers. Secondary consumers are carnivores and feed on primary consumers.

It is tempting to explain the answers to ecological problems in rather general terms and use phrases such as, 'Birds eat insects'. You must keep in mind the fact that every species has its own ecological niche and behaves in a way that is unique. So, some birds, blue tits and robins for example, do eat insects. But this is not true of all species. Sparrows and finches feed mainly on seeds, while blackbirds eat a variety of foods of which insects only make up a small proportion. In addition, blue tits and robins may feed largely on insects, but they do not eat all species. Ladybird beetles, for example, have an unpleasant bitter taste and are rarely eaten by insectivorous birds. You need to think carefully when you write about what animals and plants do.

Not all organisms behave like humans! The seeds of some plants do not germinate immediately on reaching maturity. They undergo a period of dormancy first. How is dormancy an advantage? One possible answer to this question is that dormancy means that seeds are more likely to germinate when conditions favour growth. They are not 'fooled' into growing at the wrong time of the year and dormancy does not 'save them from thinking that spring is here and germinating too early'. Birds such as starlings form flocks. This behaviour is thought to lower the risk of predation. Starlings do not form flocks because they 'know' that this will make them safer. Explaining the behaviour of different organisms in human terms is called anthropomorphism. Avoid it!

Exercise 3.1 Mark-release-recapture

Mark-release-recapture is often used to estimate the size of an animal population. This technique relies on the relationship summarised below.

$$\frac{\text{Total number of marked animals in population } [\mathbf{M}]}{\text{Total number of animals in population } [\mathbf{P}]} = \frac{\text{Number of marked animals in sample } [\mathbf{m}]}{\text{Number of animals in sample } [\mathbf{p}]}$$

1   Rearrange this expression to give a simple equation for finding the value of **P**.

(1 mark)

Blue tits are common British birds. During the winter months they form flocks and often visit gardens. When spring comes, these flocks break up. The birds disperse into woodland where they form pairs and breed. Each pair of blue tits occupies a territory in which it remains for the breeding season. Young blue tits leave their nests from late May through to the beginning of June. Parents and young remain in family groups until late summer when flocks start to form again. This information is summarised in Figure 3.1.

Month	J	F	M	A	M	J	J	A	S	O	N	D
Feeding in flocks	▓	▓	▒					▒	▓	▓	▓	▓
In breeding territories			▒	▓	▓	▓						
Young leave nest					▓	▓	▒					

Figure 3.1 *Some features of the life cycle of blue tits*

In an investigation, blue tits visiting a large garden during one day in November were trapped and marked by placing a small metal ring round one leg. A week later a second sample of blue tits was trapped. The numbers of ringed blue tits and blue tits without rings in this sample were recorded. This information is shown in Table 3.1.

Table 3.1

Number of blue tits trapped and ringed on first day	38
Number of blue tits trapped on second day which were ringed	17
Number of blue tits trapped on second day which were not ringed	8

Exercise 3.1 *continued*

2 Use the information in Table 3.1 to estimate the number of blue tits visiting the garden. Show your working.

(2 marks)

3 The mark-release-recapture technique would not give a reliable estimate of the population of blue tits in a wood in June. Give **two** reasons for this.

(2 marks)

Whales spend most of their time under water and only come to the surface to breathe. Because of this, it is very difficult to count whales accurately. It is possible, however, to approach a surfaced whale in a boat and obtain a small skin sample. In a study of humpback whales, in the Atlantic Ocean, skin samples collected in this way were used to obtain DNA profiles. These DNA profiles were then used together with mark-release-recapture to estimate the size of the humpback whale population.

4 (a) Explain what is meant by a DNA profile.

(2 marks)

(b) Explain how DNA profiles would enable researchers to estimate the size of the humpback whale population using mark-release-recapture.

(2 marks)

5 There are advantages and disadvantages with using DNA profiles rather than methods which involve marking the whales. Explain **one**

(a) advantage in using DNA profiles

(2 marks)

(b) disadvantage in using DNA profiles.

(1 mark)

The basic principle of mark-release-recapture is used in physiological as well as in ecological investigations. It has been used to estimate the volume of blood plasma in an adult human. A sample of albumin labelled with radioactive iodine, ^{131}I, was injected into a vein. After a critical time a sample of the blood plasma was withdrawn and analysed. From this the total volume of the blood plasma was estimated.

Exercise 3.1 *continued*

6 What measurements would have had to be made on the sample of blood plasma in order to estimate the total volume of blood plasma?

(2 marks)

Hint Use the expression at the start of this exercise to help you to answer this question.

7 (a) Albumin is a protein which has very large molecules. Suggest the advantage of using albumin in this investigation.

(2 marks)

(b) The albumin which was injected was labelled with radioactive iodine. Explain why it was necessary to label the albumin.

(2 marks)

8 A critical time was allowed to pass before the sample of blood plasma was withdrawn for analysis. Explain why this period of time should not have been

(a) too short

(b) too long.

(2 marks)

Exercise 3.2 Carbon dioxide and agriculture

The concentration of carbon dioxide in the atmosphere has increased considerably over the last 150 years. This is mainly as a result of burning fossil fuels.

Atmospheric carbon dioxide concentration is estimated to have been approximately 280 parts per million (ppm) in 1850. By 1989 it had risen to 353 ppm. Over the last 20 years or so, the concentration of carbon dioxide in the atmosphere has increased steadily at a rate of about 0.5% of the 1989 value per year.

1   Use the figures in the paragraph above to

      (a)   calculate the percentage of carbon dioxide in the atmosphere in 1989

(1 mark)

      (b)   estimate the percentage of carbon dioxide in the atmosphere in 1991.

(2 marks)

Show your working.

Figure 3.2 shows how an increase in carbon dioxide concentration affects the rate of photosynthesis in two important cereal crops, wheat and maize.

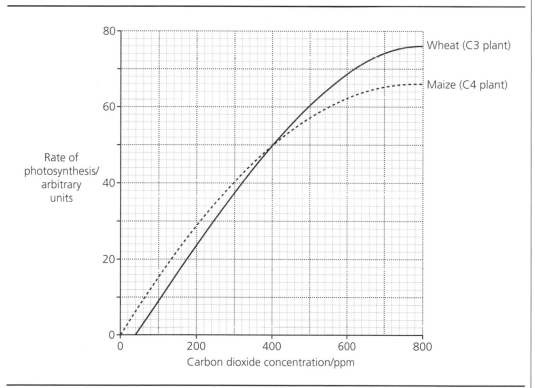

Figure 3.2 *The effect of increasing carbon dioxide concentration on the rate of photosynthesis of wheat and maize*

Different species of plant have different methods of photosynthesis. C3 photosynthesis is found in wheat and many of the plants found in temperate regions and the humid tropics. It is called C3 because the first product formed is glycerate 3-phosphate (GP) and this substance contains three carbon atoms. Maize and many plants found in the dry tropics have C4 photosynthesis.

2   Use Figure 3.2 to compare the likely effect of increasing carbon dioxide concentration from its present level on the yield of wheat with its effect on the yield of maize.

(3 marks)

3   Perennial rye-grass is a C3 plant. Maize is a C4 plant. In the UK there has been a trend to grow more maize and less perennial rye-grass as winter food for cattle. Use information from Figure 3.2 to explain whether you think this trend is likely to continue.

(2 marks)

4   Many of the world's most troublesome weeds are C4 plants growing in C3 crops.

(a)   Explain how the presence of weeds normally reduces crop yield.

(2 marks)

(b)   Use Figure 3.2 to predict the likely future effects of these weeds on crop yield. Give an explanation for your answer.

(3 marks)

An increase in atmospheric carbon dioxide concentration is thought to be one of the main causes of global warming. Table 3.2 shows how variation in mean temperature affects different stages in the growth of wheat. The data in the table refer to spring wheat. (Wheat planted in spring and harvested at the end of summer.)

Exercise 3.2 *continued*

Table 3.2 The effect of changes in mean temperature on the length of different stages in the growth of wheat

Stage in growth	Time taken in days at difference in temperature from 1990 value of			
	0°C	+1°C	+2°C	+3°C
Germination of grain	6	6	6	6
Growth of leaves	21	21	21	21
Formation of flowers	23	22	21	20
Growth of grain	22	20	18	17
Ripening of grain	12	11	10	9

5   Use Table 3.2 to calculate

   (a)   the total time from planting to harvesting at 1990 temperatures

   (1 mark)

   (b)   the percentage reduction in this time if there were a rise of 2°C in mean temperature.

   (2 marks)

Show your working.

6   Agricultural scientists have suggested that global warming will affect spring wheat production. Use the data in the table to explain how an increase in mean temperature could lead to

   (a)   a lower risk of damage by frost at the end of the growing season

   (2 marks)

   (b)   reduced yield because of the production of smaller grains.

   (2 marks)

Exercise 3.3 Controlling water hyacinths

The water hyacinth is a plant which floats on the surface of the water of lakes and slow-flowing rivers. Parts of this plant can break off very easily and grow into new plants. Because of its attractive purple flowers it was introduced into the US in the 1880s. When no native animals fed on it, it spread rapidly and soon became an important weed, choking waterways and interfering with transport and drainage.

Figure 3.3 shows the transfer of energy and nutrient cycling in an aquatic ecosystem.

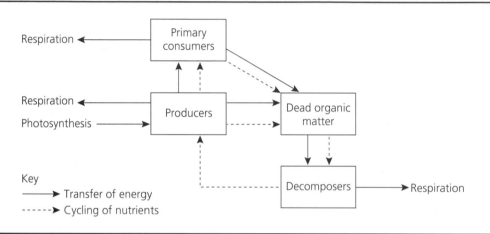

Figure 3.3 *A summary of the transfer of energy and nutrient cycling in an aquatic ecosystem*

1 Use information in Figure 3.3 to explain why the introduction of water hyacinth led to

    (a)   the accumulation of large amounts of sediment   *(2 marks)*

    (b)   a low concentration of dissolved oxygen in the water.

                                                            (2 marks)

Neochetina eichhorniae is a beetle which feeds on water hyacinths. In 1974 some of these beetles were released in an area infested with water hyacinths. The graph in Figure 3.4 shows changes in the area covered by water hyacinths following the introduction of this beetle.

2 Suggest an explanation for the change in the area of water covered by water hyacinth from

    (a)   1974 – 1975   *(2 marks)*

    (b)   1975 – 1980   *(1 mark)*

    (c)   1980 onwards.   *(2 marks)*

Exercise 3.3 *continued*

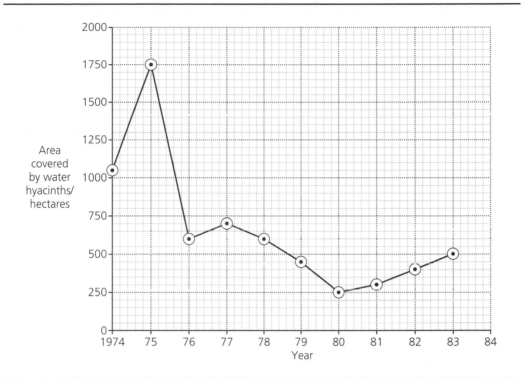

Figure 3.4 *Changes in the area of water covered by water hyacinths following introduction of the beetle,* Neochetina eichhorniae

In Zimbabwe, water hyacinths have also become a serious pest. An investigation was carried out to find out if the beetle could be used to control water hyacinths in Zimbabwe. Water hyacinths were grown individually in pots in a glasshouse. Two pairs of beetles were added to each potted plant. The graph in Figure 3.5 shows the leaf area of the plants in these pots and in control pots, at monthly intervals.

3 Two suggestions were put forward to explain the results obtained between months four and five.

- The beetles present on the plants at this time were not feeding
- The temperature in the glasshouse at this time promoted rapid growth of the water hyacinths.

Use information from Figure 3.5 to explain which of these suggestions you consider to be more likely.

(3 marks)

Table 3.3 shows some more results from this investigation.

Exercise 3.3 *continued*

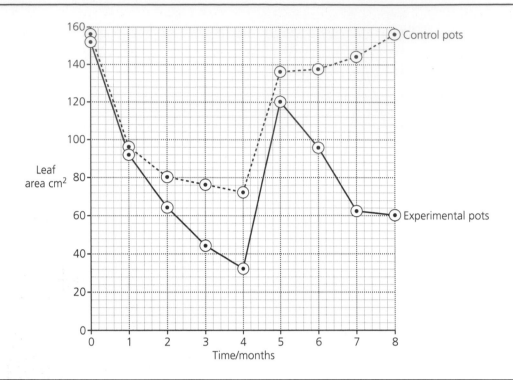

Figure 3.5 *The effect of the beetle* Neochetina eichhorniae *on leaf surface area of water hyacinth plants*

Table 3.3 Some effects of the beetle *Neochetina eichhorniae* on the growth of water hyacinths

Month	Mean number of leaves		Mean length of longest root/cm		Mean number of daughter plants	
	Exp	Control	Exp	Control	Exp	Control
1	6.5	7.1	35	26	0	0
2	6.5	6.5	10	10	6	6
3	6.5	6.5	9	12	4	6
4	7.0	6.0	7	10	5	6
5	7.0	7.0	20	19	6	6
6	7.0	7.0	21	20	4	7
7	6.0	7.0	11	20	6	7
8	6.0	6.0	11	23	7	7

4 Is *Neochetina eichhorniae* likely to be of use in controlling water hyacinths in Zimbabwe?

Use the results of this investigation to support your answer.

(3 marks)

Exercise 3.4 Competing for resources

Limpets are molluscs (shellfish) which live on rocky seashores. When the tide is in, they feed by moving slowly over the surface of rocks, grazing on algae. The graph in Figure 3.6 shows the relationship between the population density of limpets in randomly placed quadrats and their total biomass. The histograms show the distribution of shell length at different population densities.

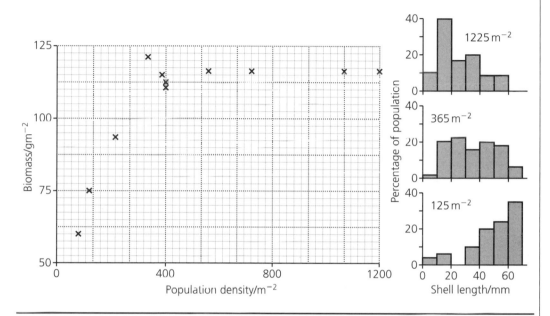

Figure 3.6 *The effect of population density on total biomass and shell length in limpets*

1   The data from which Figure 3.6 was drawn came from randomly placed quadrats. Explain why the quadrats were placed at random.

(1 mark)

2   (a)   What do the data in Figure 3.6 show about the effect of population density on limpets?

(3 marks)

(b)   Suggest an explanation for the distribution of limpet size at a population density of 1225 m⁻².

(2 marks)

Competition may affect the growth of different parts of organisms in different ways. Maize was planted at different population densities. When the plants reached maturity, their total biomass and the biomass of the grain they produced were determined. The results are shown in Table 3.4.

Table 3.4 The effect of population density on the biomass of maize

Population density of maize/plants m^{-2}	Biomass/tonnes hectare^{-1}	
	Total	Grain
7	11.1	3.8
13	11.4	2.7
25	11.8	2.1

3   The biomass of the maize plants was measured as dry mass. Explain the advantage of measuring biomass as dry mass rather than fresh mass.

(2 marks)

4   (a)   One of the resources for which the maize plants are likely to have been competing is light. Explain how plant population density will affect competition for light.

(2 marks)

   (b)   Name **two** resources other than light for which these maize plants are likely to have been competing.

(2 marks)

5   (a)   What do the results of this investigation indicate about the way in which individual maize plants grown at higher population densities differ from those grown at lower population densities?

(2 marks)

   (b)   Maize is grown for grain. It is also grown as a fodder crop to provide winter food for cattle. The cattle eat the whole plant. Use the results of this investigation to suggest advice which might be given to farmers planting maize.

(1 mark)

Exercise 3.5 Dead leaves and decomposition

Different groups of organisms are involved in breaking down the dead leaves on a woodland floor. These organisms include bacteria and fungi, arthropods such as insects and woodlice, nematode worms and earthworms. The graph in Figure 3.7 compares the biomass of these groups of organisms in a British oak woodland at different times of the year.

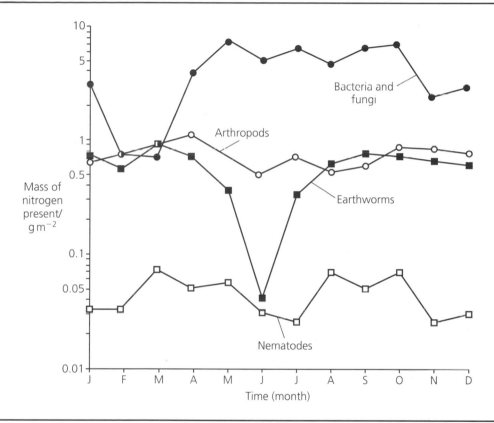

Figure 3.7 *Variation in the biomass of different groups of organisms involved in breaking down dead leaves at different times of the year*

1   In this graph, the biomass has been expressed as the mass of nitrogen present. Explain why nitrogen content can be used as a reliable measure of biomass.

(2 marks)

2   (a)   The data in the graph have been plotted on a log scale. Suggest the advantage of plotting these data on a log scale.   *(1 mark)*

(b)   Which group of organisms shows the greatest variation in biomass over the year?   *(1 mark)*

3   (a)   Calculate the ratio of bacterial and fungal biomass to arthropod biomass in June.   *(2 marks)*

(b)   Suggest an explanation for the change in biomass of bacteria and fungi between February and June.

(3 marks)

The importance of arthropods and microorganisms in the decomposition of leaves was investigated. Different numbers of woodlice were added to containers, each with the same mass of oak leaves. The rates of respiration of the microorganisms in these containers and in a control container were measured. The results are shown in Figure 3.8.

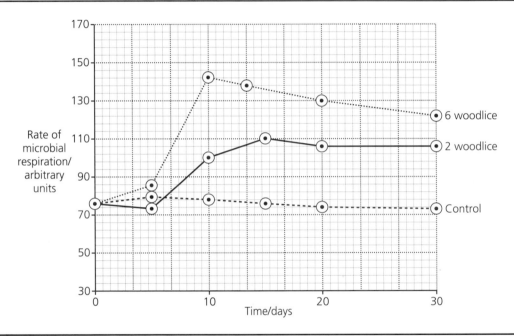

Figure 3.8 *The effect of woodlouse activity on microbial breakdown of leaves*

4   Describe how the control container should have been set up.

(2 marks)

5   (a)   What does Figure 3.8 show about the effect of woodlouse activity on the microbial breakdown of leaves?

(2 marks)

(b)   Explain **one** way in which the presence of woodlice could have caused the change in the microbial respiration rate, which occured in the first 7 days.

(2 marks)

Exercise 3.6 Seaweeds on the seashore

Seaweeds photosynthesise. They are algae and belong to the kingdom Protoctista. Many species grow in the intertidal zone on rocky seashores. Figure 3.9 shows the distribution of some common seaweeds.

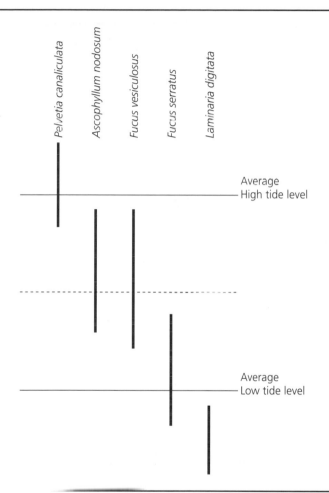

Figure 3.9 *The zonation of some common species of seaweed on a rocky seashore*

1   Briefly describe how you would

    (a)   position a transect to obtain data on the distribution of seaweeds

                 (1 mark)

    (b)   use a point quadrat frame to obtain quantitative data on seaweed distribution.

                 (2 marks)

Exercise 3.6 *continued*

2 On seashores that are exposed to larger waves, most species of seaweed extend farther up the shore than they do on sheltered seashores. Suggest why.

(1 mark)

The complete tide cycle takes about twelve and a half hours. This is the time taken for the sea to fall from high tide level to low tide level and rise back again to high tide level.

3 The environmental conditions are different on the area of seashore where *Pelvetia canaliculata* occurs and on the area occupied by *Laminaria digitata*. Explain how they differ with respect to:

    (a)   the length of time covered by seawater

(1 mark)

    (b)   variation in temperature

(2 marks)

    (c)   variation in the salt content of the water in contact with the seaweed.

(3 marks)

Table 3.5 shows data concerning some of the seaweeds whose distribution is shown in Figure 3.9.

Table 3.5 Some data concerning seaweeds growing on a rocky shore

Species	Thickness of cell wall when seaweed is fully hydrated/μm	Lipid content of cell wall/%	Loss in mass of seaweed when exposed to air for 12 hours/%
Pelvetia canaliculata	1.48	4.88	50
Ascophyllum nodosum	1.01	2.87	62
Fucus vesiculosus	0.68	2.60	64
Fucus serratus	0.43	n/a	72

4 Describe how position on the seashore is related to the data shown in Table 3.5.

(1 mark)

5 Use the data to explain how *Pelvetia canaliculata* is adapted to living on the upper part of the seashore.

(4 marks)

Exercise 3.7 Insecticides, eggshells and environments

A good insecticide has three properties. It must be toxic or it will not kill its target organisms. It must be persistent so that its effects will last. And finally, it must not harm other organisms. In the early 1950s, DDT was thought to have all these properties and its use became widespread. Coinciding with this, however, came a sharp decline in the numbers of some species of bird. One of these was the peregrine falcon, a bird of prey which feeds mainly on other birds.

Research showed that, far from being harmless, DDT had a number of damaging effects on peregrine falcons. One of these was on eggshell thickness. In an investigation, shell thickness and strength were measured on two samples of eggs. The first sample was obtained from museum collections and represented eggs laid between 1850 and 1942, before the use of DDT. A second sample, collected in the wild, was from peregrines that had been exposed to DDT. The graph in Figure 3.10 shows the results of this investigation.

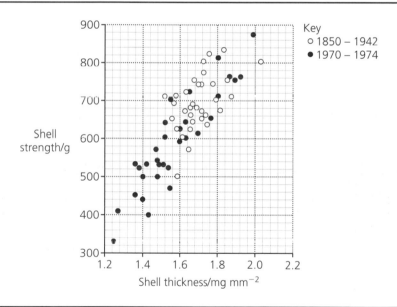

Figure 3.10 *Shell thickness and strength in eggs of peregrine falcons laid from 1850 to 1942 and from 1970 to 1974*

1   (a)   Shell strength was measured by determining the weight required to push a needle through the shell, under standard conditions. Suggest **one** condition which would need to be standardised in making these measurements. Give a reason for your choice.

(2 marks)

(b)   The museum eggs had had their contents removed so that only shells remained. Explain the advantage of measuring their thickness in $mg\,mm^{-2}$.

(1 mark)

2   The differences between the two groups of eggs were found to be significant at the 0.05 probability level. Explain the meaning of *significant* at the 0.05 probability level.

(2 marks)

3   It was suggested that there was a link between shell thickness and hatching success of the eggs. Describe and explain this link.

(2 marks)

Fish such as minnows are also affected by DDT. In another investigation, minnows were divided into four groups which were treated as follows.

Group **A**   exposed to DDT in the water

Group **B**   fed on clams containing DDT

Group **C**   exposed to DDT in the water and fed on clams containing DDT

Group **D**   control group

Exercise 3.7 *continued*

The concentration of DDT in the bodies of these minnows was measured at various times. Some of the results of this investigation are shown in Figure 3.11

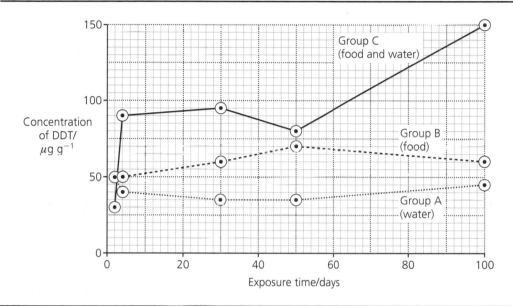

Figure 3.11 *The concentration of DDT in the bodies of minnows exposed to DDT, in the water in which they were swimming and in their food*

4   Explain why a control group is necessary in this investigation.

(1 mark)

5   What conclusions can be drawn from Figure 3.11 about the ways in which DDT enters the body tissues of a fish such as a minnow?

(3 marks)

6   (a)   The mean concentration of DDT in the clams fed to the minnows was 1.8 mg g^{-1}. By how many times was the DDT from the clams concentrated in the tissues of the minnows after 100 days?

(1 mark)

(b)   Explain why the concentration of DDT in minnows is higher than the concentration in their food.

(2 marks)

(c)   Use the results of this investigation to suggest why peregrine falcons are particularly susceptible to poisoning by DDT.

(1 mark)

CHAPTER 3

Exercise 3.8 Tigers and their prey

Tigers are, or were once, distributed over much of Asia. In the past hundred years their numbers have declined rapidly, largely as a result of human activity. In many parts of their range they are now facing extinction. One of the factors which governs whether or not a particular area can support a healthy tiger population is the availability of prey.

We can explore the relationship between tiger and prey numbers by looking at a simplified situation where tigers are feeding on a single species of prey, a small deer called a muntjac. Table 3.6 contains some data about muntjac and the feeding habits of tigers.

Table 3.6 The food requirements of tigers

Muntjac	
Body mass/kg	20
Population density of adults/ number 100 km^{-2}	500
Tiger	
Food requirements of a female for maintenance/kg meat day^{-1}	6

1   A pair of muntjacs produce 1 young per year. Calculate the muntjac population after one year, assuming that no muntjac died.

(1 mark)

2   (a)   30% of a muntjac cannot be eaten. Calculate the number of muntjac a female tiger on a maintenance diet would require to kill in a year. Show your working.

(2 marks)

   (b)   A pregnant tiger or one feeding young requires 50% more food. How often would a female tiger have to kill a muntjac if she were feeding young? Show your working.

(2 marks)

3   (a)   Gunung Leuser National Park is in Sumatra. It has a high rainfall and is covered in thick forest. The only species of ground-feeding herbivorous mammal present is the muntjac and it has a low population density. Suggest an explanation for the low population density of ground feeding herbivores in an area of thick forest.

(2 marks)

Exercise 3.8 *continued*

(b)   The territories over which female tigers hunt in Gunung Leuser are very large. Use information given in this exercise to explain why.

(2 marks)

The graph in Figure 3.12 shows the relationship between the population densities of three species of carnivore and prey biomass.

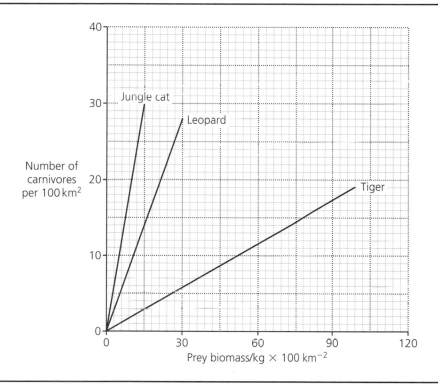

Figure 3.12 *The relationship between carnivore population density and prey biomass for three species of cat found in India. These species differ in size. The jungle cat has a body mass of 4 – 10 kg, the leopard 30 – 90 kg and the tiger 150 – 220 kg*

4 In some areas, the numbers of some wild herbivorous mammals have been reduced considerably by poaching. Use Figure 3.12 to describe how this would affect the population densities of these three carnivores in the area concerned.

(2 marks)

Research has been carried out on the prey of tigers in Panna Tiger Reserve in India. This reserve is surrounded by a large number of villages. The people in these villages are farmers and keep cattle. The researchers collected tiger faeces over a period of a year and then analysed random samples collected in three different parts of the year: the cool dry winter months, the hot dry months of early summer and the monsoon (a period of heavy rain towards the end of the summer). The results are shown in Table 3.7.

Exercise 3.8 *continued*

Table 3.7 The prey of tigers in Panna Tiger Reserve at different time of the year

Species	Percentage of tiger faecal samples containing hair from this species in		
	cool dry season	hot dry season	monsoon
Large deer and antelope			
Sambar	13.3	46.7	13.3
Nilgai	40.0	13.3	20.0
Chital	13.3	0	40.0
Small antelope			
Chousingha	13.3	0	6.7
Other wild mammals			
Langur monkey	0	20.0	6.7
Porcupine	13.3	0	6.7
Wild pig	13.3	6.7	6.7
Domesticated mammals			
Cattle	13.3	20.0	0

5   The figures in Table 3.7 are expressed as percentages.

    (a)   The figures for the dry season add up to more than 100. Explain why.

           (1 mark)

    (b)   What do the figures in Table 3.7 suggest about the reliability of this research? Explain your answer.

           (2 marks)

6   Suggest explanations for the percentage of samples from faeces produced in the hot dry season containing hair from:

    (a)   langur monkeys   *(2 marks)*

    (b)   cattle.   *(2 marks)*

7   Using any of the data provided in this exercise, suggest why it is particularly difficult to establish reserves and National Parks to conserve tiger populations in countries like India.

           (2 marks)

CHAPTER

four | **Genes and genetics**

If you can think clearly and logically you ought to welcome genetics problems. You will inevitably encounter such problems somewhere in your A Level examinations but they are usually straightforward if you follow the necessary conventions and observe a few basic rules. We will illustrate these conventions with a problem involving a dihybrid cross. This is a cross which involves two different genes.

In peas, the height of the plant is determined by a gene. The allele for tall plants, **T**, is dominant to the allele for dwarf plants, **t**. A second gene, situated on another chromosome, determines pod colour. The allele for green pods, **G**, is dominant to the allele for yellow pods, **g**. A cross was carried out between homozygous tall plants with green pods and homozygous dwarf plants with yellow pods. All the F_1 plants which resulted from this cross were tall with green pods. Use a genetic diagram to show the expected results of a cross between two of the F_1 plants.

The key to success with any genetics problem is to follow the layout shown on page 62. It is what your examiners will expect and it is the approach that they will follow. Note, in particular, the points which have been picked out. Exercises 4.1 to 4.6 in this chapter are all genetics problems and should be approached in this way, where appropriate. They are, however, a little different from the straightforward problems with which you may be familiar. You will need to draw on your data-handling skills as well as on other aspects of your biological knowledge.

Set out your method as shown in this diagram. The approach used in this example is recommended by all examination boards, Use it as a model for your answers.

When you are writing quickly, it is easy to confuse capital and small letters. Try to avoid using letters such as C and c and S and s to represent alleles.

Parental phenotypes   Tall, green pods   Tall, green pods

Parental genotypes: TtGg   TtGg

Gametes: (TG) (Tg) (tG) (tg)   (TG) (Tg) (tG) (tg)

Offspring genotypes:

Use a Punnett square to work out the cross. You are not so likely to make mistakes this way. Whatever you do, avoid linking gametes to the relevant genotypes. This can get very confusing.

Draw a ring round each gamete. Remember that, since gametes are haploid, you can only have one pair of alleles in each ring.

	Male gametes			
	(TG)	(Tg)	(tG)	(tg)
(TG)	TTGG	TTGg	TtGG	TtGg
(Tg)	TTGg	TTgg	TtGg	Ttgg
(tG)	TtGG	TtGg	ttGG	ttGg
(tg)	TtGg	Ttgg	ttgG	ttgg

Female gametes

Offspring phenotypes
Tall, green pods	9
Tall, yellow pods	3
Dwarf, green pods	3
Dwarf, yellow pods	1

Exercise 4.1 Chlorophyll and wheat

In wheat, the flag-leaf is the last leaf to be produced. It is on the stalk which carries the grain. The concentration of chlorophyll in the flag-leaf is controlled by a single gene. The allele for high chlorophyll concentration, **H**, is dominant to that for low chlorophyll concentration, **h**.

Plants with the genotypes **HH** and **hh** were crossed to produce an F_1 generation. The plants in the F_1 generation were then interbred to produce an F_2 generation. Figure 4.1 shows the chlorophyll concentration in the flag-leaves of the parent plants, the F_1 plants and the F_2 plants.

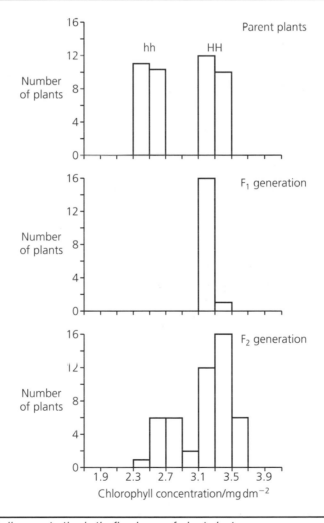

Figure 4.1 *Chlorophyll concentration in the flag-leaves of wheat plants*

Exercise 4.1 *continued*

1 Explain the evidence from Figure 4.1 that

  (a) the allele for high chlorophyll concentration is dominant to that for low chlorophyll concentration

  (1 mark)

  (b) the environment may affect the chlorophyll concentration in the flag-leaf.

  (1 mark)

2 Explain the results obtained for the F_2 generation.

  (2 marks)

3 Some F_1 plants were crossed with recessive homozygotes, **hh**. Describe the shape of the histogram you would expect if you measured the chlorophyll content of the flag-leaves of the plants resulting from this cross.

  (1 mark)

Exercise 4.2 Tuberculosis and genetics

Tuberculosis (TB) is an infectious disease which often affects the lungs. It is caused by a bacterium. Table 4.1 shows the probability of one of two close relatives catching tuberculosis when the other already has the disease.

Table 4.1 Probability of second relative catching tuberculosis in different relationships

Relationship	Probability of second relative catching tuberculosis
Identical twins	0.87
Non-identical twins	0.26
Full siblings	0.26
Half siblings	0.12
Parent and offspring	0.17
Husband and wife	0.07

- Full siblings are brothers and sisters. Half-siblings are half brothers and half sisters. They have either the same father or the same mother as each other.
- The probability of catching tuberculosis from a member of the general population was estimated to be 0.01.

1 Give **two** different hypotheses to suggest why relatives might both be expected to get tuberculosis. Name one hypothesis **A** and the other, hypothesis **B**.

(2 marks)

2 For each hypothesis, select two values from the data to show a contrast which supports the hypothesis. In each case, explain how these values provide support.

(a) Present values from the data and your explanation for hypothesis **A**.

(2 marks)

(b) Present values from the data and your explanation for hypothesis **B**.

(2 marks)

Exercise 4.2 *continued*

Isoniazid is a drug used to treat tuberculosis. The histogram in Figure 4.2 shows the concentration of isoniazid in the blood plasma of a group of people. The results of this investigation were obtained 6 hours after the people concerned were given a standard dose of the drug.

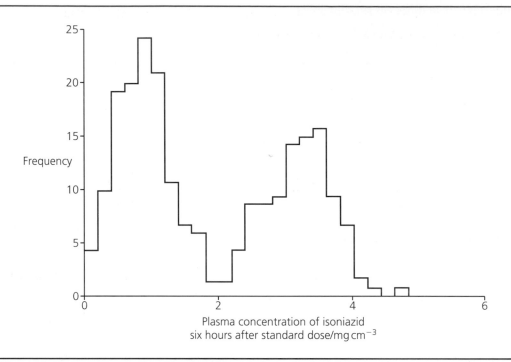

Figure 4.2 *Plasma concentration of isoniazid 6 hours after a standard dose*

3 Explain why it was necessary to give different amounts of the drug to different people in order to give a standard dose of isoniazid.

(1 mark)

4   (a)   Describe the relationship between the concentration of isoniazid in the blood plasma and the rate at which the drug is excreted from the body.

(1 mark)

(b)   Explain the evidence from Figure 4.2 that the ability to excrete isoniazid is controlled by a single pair of alleles, one recessive to the other.

(2 marks)

Exercise 4.3 Sex determination

In mammals, X and Y chromosomes determine the sex of an individual. In other organisms, sex may be determined in different ways.

A queen honeybee can lay both fertilised and unfertilised eggs. Fertilised eggs develop into females which are always diploid. Unfertilised eggs develop into males. Males are, therefore, always haploid. Figure 4.3 shows the formation of gametes in male and female honeybees.

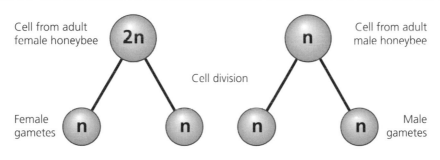

Figure 4.3 *Gamete formation in honeybees*

1 Giving a reason for your choice, name the type of cell division which produces male gametes.

(2 marks)

2 Body colour in honeybees is determined by a single gene. The allele, **B**, for yellow body is dominant to the allele, **b**, for black body. Explain why, in a mating between a homozygous black female and a yellow male,

(a) all the female offspring were yellow

(2 marks)

(b) all the male offspring were black.

(1 mark)

Exercise 4.3 *continued*

In the squirting cucumber, there are three kinds of plants: plants with male flowers only, plants with female flowers only and hermaphroditic plants with both male and female flowers. Sex is determined by a single gene with three alleles.

A^D is the allele for male. It is dominant over the other two alleles.

A^+ is the allele for hermaphrodite. It is dominant over allele A^d.

A^d is the allele for female. It is recessive to the other two alleles.

3   (a)   Give the genotype of a female squirting cucumber.

(1 mark)

(b)   Explain why a male squirting cucumber cannot have the genotype $A^D A^D$.

(2 marks)

4   Draw a genetic diagram to show how it would be possible for a cross between two hermaphroditic plants to produce female offspring.

(2 marks)

Exercise 4.4 Saddleback pigs

Pigs are farmed for meat. Clearly, the more piglets a sow produces, the more profit a farmer should be able to make. A sow is only fertile around the time she ovulates. At this time, she shows a distinctive pattern of behaviour known as oestrus. She is willing to mate and is likely to become pregnant. An experienced pig farmer can recognise oestrus and can use this knowledge either to allow a boar access to her or to inseminate her artificially. This sounds an ideal way of ensuring that the sow will become pregnant, but there is a practical problem. Mating must take place at the right time during oestrus or fertilisation is less likely. Figure 4.4 shows the relationship between the time at which a sow mates and the percentage of eggs that are fertilised.

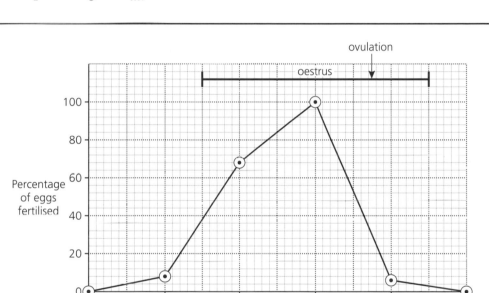

Figure 4.4 *The relationship between the time of mating and fertilisation in pigs*

1 (a) Use Figure 4.4 to suggest the maximum time that sperm cells remain viable inside the reproductive system of a sow. Explain how you arrived at your answer.

(2 marks)

(b) If mating takes place 60 hours after the start of oestrus, very few eggs are fertilised. Suggest why and explain how you arrived at your answer.

(2 marks)

Exercise 4.4 *continued*

In pigs, the allele for white is dominant to the allele for saddleback. A saddleback sow was mated with a saddleback boar 12 hours after the start of oestrus. She was mated again, this time with a white boar, 48 hours after the start of oestrus. She produced 12 young, 10 of which were saddleback and 2 of which were white. All three adult pigs were homozygous.

2   What would you expect to be the appearance of the piglets produced by the saddleback female if the white boar had been the father? Use a genetic diagram to explain your answer.

(2 marks)

3   One conclusion that might be drawn from the result of this double mating is that mating is more likely to result in fertilisation if it takes place 12 hours after the start of oestrus than if it takes place after 48 hours. Explain

(a)   how the results of the matings support this conclusion

(2 marks)

(b)   why this conclusion may not be reliable.

(2 marks)

Exercise 4.5 Sterile cherries and fertile bananas

In flowering plants, pollen grains are produced by meiosis and have a haploid number of chromosomes. Pollination involves the transfer of pollen to the stigma of the female parent. The pollen grain germinates and a pollen tube grows down the stigma and style into the ovary where fertilisation takes place.

Research carried out with cherries showed that when trees were self-pollinated, they did not produce fruit. The explanation was found to be genetic and controlled by a gene, S, with a large number of different alleles S_1, S_2, S_3, S_4, S_5 etc. A pollen grain carrying a particular allele will not grow on a style which contains the same allele. Table 4.2 shows the genotypes of some different cherry varieties.

Table 4.2 The genotypes of some different varieties of cherry

Variety	Genotype
Bedford Prolific	S_1S_2
Early Rivers	S_1S_2
Belle Agathe	S_1S_3
Frogmore Early	S_1S_3
Bigarreau Napoleon	S_3S_5
Governor Wood	S_1S_4

1 For a Bedford Prolific, give the genotype or genotypes of

   (a)   a pollen grain

   (b)   a cell from the stigma

(1 mark)

2 Use your answer to Question 1 to explain why Bedford Prolific does not produce fruit when self-pollinated.

(2 marks)

3 With which of the varieties listed in Table 4.2 would you pollinate a Bedford Prolific if you wanted to produce the largest crop of cherries. Explain your answer.

(2 marks)

Exercise 4.5 *continued*

4 The variety Early Rivers was crossed with Governor Wood. Early Rivers was the female parent. What would be the genotypes of the offspring produced by this cross? Use a genetic diagram to explain your answer.

(2 marks)

Cultivated bananas are usually sterile and banana fruits do not contain seeds. This creates difficulties for plant breeders trying to produce new varieties of banana. One of the most widespread banana varieties is Gros Michel. This variety is triploid, each of its cells containing 33 chromosomes.

5 Use your knowledge of cell division to suggest why Gros Michel bananas do not produce fertile seeds.

(3 marks)

Breeding programmes used to produce new varieties of banana, rely on the fact that Gros Michel occasionally produces fertile female gametes. These gametes contain 33 chromosomes. The diagram in Figure 4.5 shows a scheme for producing new varieties of banana by breeding.

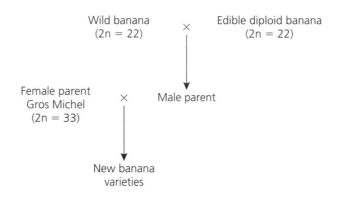

Figure 4.5 *Breeding new banana varieties*

6 How many chromosomes are there in a leaf cell from one of the new banana varieties resulting from this scheme? Explain how you arrived at your answer.

(2 marks)

Exercise 4.5 *continued*

7 The new banana varieties were fertile and were very similar to Gros Michel. When they were crossed with each other, their offspring were found to vary considerably and showed few of the qualities of Gros Michel. Suggest an explanation for each of the following

(a)   the new banana varieties were fertile

(1 mark)

(b)   the new banana varieties were similar to Gros Michel

(1 mark)

(c)   the characteristics of the offspring resulting from crossing the new banana varieties were variable.

(1 mark)

Exercise 4.6 Rats, Warfarin and heart disease

Before blood clots at the site of an injury, a number of different proteins must be present in the plasma. These proteins are called clotting factors and many of them are synthesised in the liver. One of the reactions necessary for their synthesis is the production of an unusual amino acid called γ-carboxyglutamic acid. This reaction requires vitamin K.

1 Vitamin K is not transferred very efficiently across the placenta. Suggest how this may affect new-born babies.

(1 mark)

Warfarin is a substance which has been used very successfully as a rat poison. It acts as a competitive inhibitor of vitamin K. In 1959, rats resistant to Warfarin were discovered in an area of central Wales. These rats were investigated and it was found that resistance was determined by a single gene with codominant alleles, R^S and R^R. Table 4.3 shows the characteristics of rats with the different genotypes which result from these alleles.

Table 4.3 The characteristics of rats with the genotypes R^SR^S, R^SR^R and R^RR^R

Genotype	Susceptibility to Warfarin	Vitamin K requirement
R^SR^S	Susceptible	Require small amounts in diet
R^SR^R	Resistant	Require more than rats with genotype R^SR^S
R^RR^R	Resistant	Require very large amounts and cannot survive under natural conditions

2 Giving an explanation for your answer in each case, which of the genotypes in Table 4.3 would be at the greatest advantage in an area where

(a) no Warfarin was used

(1 mark)

(b) Warfarin was used extensively as a rat poison.

(1 mark)

3 Use suitable genetic diagrams to show the expected genotypes of the offspring of crosses between rats with the genotypes

(a) R^SR^S and R^SR^S

(b) R^SR^R and R^SR^R

(3 marks)

Exercise 4.6 *continued*

4   Figure 4.6 shows the change in frequency of Warfarin-resistant rats in central Wales in the years after Warfarin was first used as a rat poison.

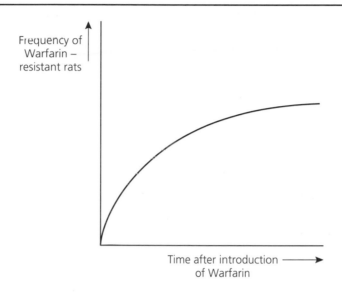

Figure 4.6 *The change in frequency of Warfarin-resistant rats following the introduction of Warfarin*

Use the results of the crosses in Question 3 and the information provided to explain the shape of the curve in Figure 4.6.

(4 marks)

Warfarin is also used in medicine as an anticoagulant. It is often used on patients with deep-vein thrombosis, where a blood clot forms, usually in the veins of the legs.

5   If a patient who is taking Warfarin bleeds, he or she is given vitamin K. Explain why this treatment is successful.

(1 mark)

6   Heparin is another drug used as an anticoagulant. It inhibits the activity of the enzyme which converts prothrombin to thrombin. Explain why heparin is quicker acting than Warfarin.

(2 marks)

7   Humans possess a gene for Warfarin resistance similar to that in rats. Suggest why extensive use of Warfarin to treat cardio-vascular disease would be unlikely to change the frequency of Warfarin resistance in the human population.

(2 marks)

Exercise 4.7 Living in caves

The freshwater shrimp *Gammarus minus*, shown in Figure 4.7, is a small crustacean that lives in streams. It feeds on detritus and microorganisms which it filters from the water. It is eaten by several species of fish.

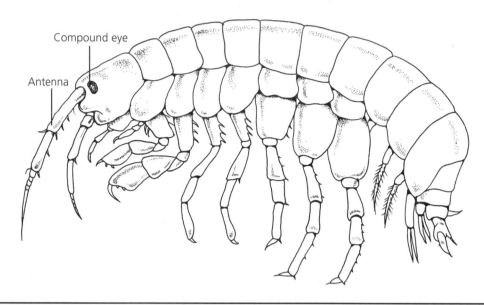

Compound eye

Antenna

Figure 4.7 *The main features of* Gammarus minus

Some populations of *G. minus* live in streams in the open and some live in total darkness in streams which run through caves. The freshwater shrimps living in these two environments differ from each other in several ways. In one investigation, the lengths of the body and of the antennae were measured in different populations of *G. minus*. For each population mean body length and mean antenna length were calculated. The results of the investigation are shown in Figure 4.8.

Exercise 4.7 *continued*

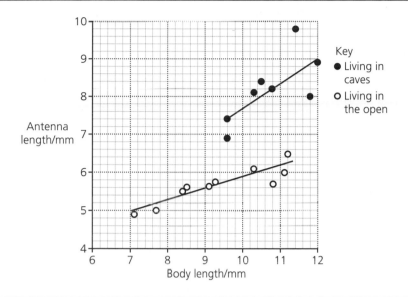

Figure 4.8 *Mean antenna length and mean body length of populations of G. minus living in the open and in caves*

1 What information does the graph give about body length and antenna length of the shrimps living in the two environments?

(3 marks)

2 In a further study of these shrimps, the parts of the central nervous system associated with chemical stimuli and with visual stimuli were compared. The relevant parts were measured and the results were tested statistically. It was found that the part associated with chemical stimuli was significantly larger in the cave-dwelling shrimps while the part associated with visual stimuli was significantly smaller.

(a) The part of the central nervous system associated with chemical stimuli was significantly larger in cave-dwelling shrimps. Explain the meaning of *significantly*.

(2 marks)

(b) Suggest how long antennae may have a selective advantage for shrimps which live in caves.

(3 marks)

Freshwater shrimps have compound eyes. Each of these eyes is made up of a number of ommatidia. An ommatidium consists of a lens system and a group of sense cells. The responses of freshwater shrimps from caves and from springs out in the open were investigated. The shrimps were placed in tubes which offered a choice of light or dark conditions. Some of the results of this investigation are shown in Table 4.4.

Table 4.4 The response of freshwater shrimps, from different populations, to light

Population	Mean number of ommatidia per eye	Percentage of time spent in the light	N
Davis spring	28	17.5	38
Organ spring	25	21.1	38
Organ cave	3	40.4	27
Benedict's cave	4	32.5	37

3   In the table, **N** represents the number of shrimps tested in each population. Explain why it is useful to know the value of **N**.

(1 mark)

4   If light made no difference to the behaviour of the shrimps, what percentage of time would you expect the animals to spend in the light conditions? Explain your answer.

(2 marks)

5   (a)   Explain the advantage of the response to light demonstrated by shrimps from the spring populations.

(2 marks)

        (b)   Describe and suggest an explanation for the difference in behaviour of the cave-dwelling shrimps.

(2 marks)

Exercise 4.8 Bacteria, plasmids and antibiotic resistance

Many diseases caused by bacteria are treated with antibiotics. Some of these antibiotics are losing their effectiveness, however, because bacteria are becoming resistant to them. The proportion of antibiotic-resistant bacterial cells in a population increases as a result of selection.

1   (a)   How do bacteria resistant to a particular antibiotic arise in a population?

(1 mark)

        (b)   Explain how the use of antibiotics could lead to an increase in the proportion of antibiotic-resistant bacteria in a population.

(3 marks)

2   Most bacteria reproduce asexually by the process of binary fission. This involves the DNA in their single 'chromosome' replicating and the cytoplasm dividing to produce two identical cells. Use this information to explain why genes for antibiotic resistance spread through generations in bacterial populations more rapidly than do genes in human populations.

(3 marks)

The DNA in a bacterial 'chromosome' carries the genes which code for essential proteins. Plasmids are extra pieces of DNA which exist and are able to divide independently of the bacterial 'chromosome'. They often contain genes that code for proteins which are only required in abnormal situations. Some of these genes code for antibiotic resistance and allow bacteria to survive in the presence of antibiotics. Plasmids also possess genes which enable bacteria to pass copies of plasmids to other bacterial cells. These bacteria may even belong to a different species.

Transfer of plasmids may be demonstrated experimentally. A donor strain of bacterial cells has plasmids which make these cells resistant to streptomycin and penicillin. They are, however, sensitive to tetracycline. This antibiotic kills them. A recipient strain is resistant to tetracycline because of a mutation in its 'chromosomal' DNA but these recipient bacteria are sensitive to penicillin and streptomycin. This information is summarised in Table 4.5.

Table 4.5 Characteristics of donor and recipient bacteria used to demonstrate transfer of plasmids

Bacterial strain	Resistant to		
	penicillin	streptomycin	tetracycline
Donor	✔	✔	✗
Recipient	✗	✗	✔

Key
✔ resistant
✗ sensitive

3   (a)   Cultures of the two strains were mixed and a sample of the mixture removed immediately. This sample was spread over an agar plate containing the antibiotics penicillin, streptomycin and tetracycline. No bacterial colonies developed. Explain why.

(2 marks)

(b)   How would you use the agar plate described in part (a) to distinguish between recipient bacteria which had acquired resistance and those which had not?

(1 mark)

The graph in Figure 4.9 shows the results of this experiment. The curve levels out when all the recipient bacteria which are capable of acquiring plasmids have done so.

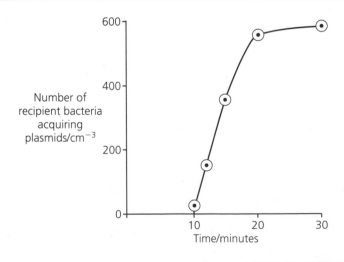

Figure 4.9 *The spread of antibiotic resistance plasmids in a population of recipient bacteria*

Exercise 4.8 *continued*

4 What does the graph suggest about the time taken for a plasmid to be transferred from one bacterial cell to another? Give the reason for your answer.

(2 marks)

5 This experiment was repeated. The number of bacteria in the original mixture was kept the same, but the ratio of donor:recipient bacteria was reduced. Describe how this would affect the curve showing the results of the experiment.

(2 marks)

In another investigation, the effects of different concentrations of two antibiotics on bacteria containing resistance plasmids were investigated. Table 4.6 shows the results.

Table 4.6 The minimum inhibitory dose of the antibiotics streptomycin and ampicillin for bacteria containing resistance plasmids

Species of bacterium	Notes	Minimum inhibitory dose/μg cm^{-3}	
		streptomycin	ampicillin
Escherichia coli	Found in gut. Often harmless but some strains cause diarrhoea	20	25
Shigella sonnei	Causes mild diarrhoea in children. Common in UK	200	250
Salmonella typhimurium	A common cause of food poisoning the UK	125	100
Proteus mirabilis	Not normally harmful but can cause diarrhoea when large numbers ingested	100	25
Klebsiella pneumoniae	Causes respiratory infections but commonly found in the gut where harmless	125	250

Note   When no resistance plasmids were present, the minimum inhibitory dose for streptomycin was between 0.3 and 10 μg cm^{-3}. For ampicillin it was between 3 and 7.5 μg cm^{-3}.

6 Give **two** general conclusions that you can draw from the data in Table 4.6.

(2 marks)

7 (a) Explain how the use of antibiotic-resistant genes enable genetically-modified bacteria to be detected.

(2 marks)

   (b) Use information given in this exercise to suggest why care should be taken in using resistance genes in this way.

(2 marks)

Exercise 4.9 Our closest ancestor

Our current view of human evolution is that humans separated from an ape-like ancestor around five million years ago. Fossil evidence for this is very limited and, in trying to re-construct our evolutionary past, we rely to a large extent on comparing present-day humans, gorillas and chimpanzees. These animals are very similar to each other and, almost certainly, shared a common ancestor. The diagram in Figure 4.10 shows some possible evolutionary pathways.

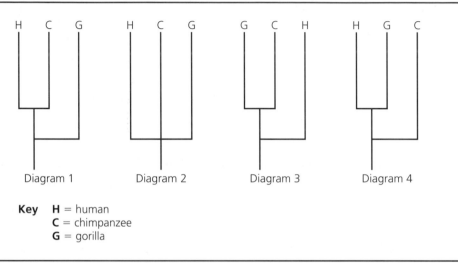

Key   H = human
   C = chimpanzee
   G = gorilla

Figure 4.10 *Possible evolutionary pathways involving humans, gorillas and chimpanzees*

1   Diagram **1** can be interpreted as suggesting that humans and chimpanzees are most closely related and that their lines separated most recently. The gorilla line split away earlier. Suggest an interpretation for Diagram **2**.

(1 mark)

2   Although a single piece of evidence may support a particular interpretation, other interpretations may be possible. Illustrate this statement by referring to chromosome numbers from Table 4.7.

(2 marks)

Exercise 4.9 *continued*

Table 4.7 Chromosome numbers in some species of primate

Species	Chromosome number
Human	46
Chimpanzee	48
Gorilla	48
Orang-utan	48
Baboon	42
White-handed gibbon	42

Albumin is a protein found in the blood plasma. It consists of a single polypeptide chain, 584 amino acids in length.

3   The sequence of amino acids in the albumin from different species differs slightly. Explain what causes these slight differences in amino acid sequence.

(2 marks)

In an investigation, a comparison was made between the albumin of different species. A rabbit was injected with human albumin. As a result, it produced human albumin antibodies in its blood. There were 30–40 different antibodies produced, each specific to a particular antigenic site on the human albumin molecule. A sample of human albumin was mixed with these antibodies. The antibodies bound to the albumin and formed large particles which precipitated from the solution.

4   Use your knowledge of protein structure to explain why the antibodies produced by the rabbit bound to human albumin.

(2 marks)

Another sample of the antibodies obtained from the rabbit was mixed with baboon albumin. The amino acid sequence of baboon albumin differs slightly from that of human albumin. As a result, the antibodies bound at fewer sites and less precipitate was formed. Table 4.8 shows the amount of precipitate formed when human albumin antibodies were mixed with albumin from a number of different species.

Exercise 4.9 *continued*

Table 4.8 The antigen-antibody reaction between albumin and human albumin antibodies in different species

Species tested	Reaction (as a percentage of that with human albumin)
Human	100
Chimpanzee	95
Gorilla	95
Orang-utan	85
Gibbon	82
Baboon	73
Kangaroo	8

5 Explain what the figures in Table 4.8 suggest about the evolutionary relationship between baboons and kangaroos.

(2 marks)

6 (a) It has been suggested that the data in Table 4.8 provide evidence for the evolutionary pathway shown in Diagram **3** in Figure 4.10. Explain how.

(1 mark)

(b) What other antibody-antigen test would you need to carry out to support this pathway? Explain your answer.

(2 marks)

7 The differences found in this investigation are due to differences in the sequences of amino acids in the different albumin molecules. Explain why this test could be made more sensitive by looking at differences in the DNA base sequence in the albumin gene rather than differences in the amino acid sequence.

(3 marks)

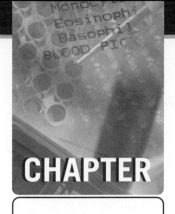

Organs and systems

The exercises in this chapter involve the physiology of living organisms. This means the way in which their organs and systems function. Physiology is one of the more difficult areas of biology and physiological topics can be difficult to understand. If you look in a library, it is more than likely that you will find some old biology textbooks. Browse through these and you will see that much emphasis was placed on describing what organs looked like and how they were arranged in the body. Compare this with the approach in the textbook you are using. Here the emphasis is placed more on the way in which systems work – and this is often explained at the molecular and cellular levels. It is not surprising that physiology can be difficult! However there are two things which, if you bear them in mind, should help to ensure a successful approach to the exercises in this chapter.

Different systems work together

From a teaching and possibly a learning point of view, it is more straightforward to look at each system in the body of a mammal independently. In practice though, all these systems work together. Think, for example, about the digestive system. Its main function is the breakdown of large insoluble molecules in the food of an organism into smaller, soluble ones and the absorption of these molecules through the wall of the intestine. But, to do this, other systems are involved. The gut wall is made of living cells and these cells need oxygen in order to respire, so the gas-exchange and blood systems are linked to the functioning of the gut. Muscle contraction results in the movement of substances along the intestine and these muscles are controlled in part by nerve impulses. A real understanding of the functioning of the digestive system clearly relies on an understanding of other systems in the body. When you encounter a problem centred on a particular system, do not forget about all the others!

An understanding of physiology relies on basic biological principles

Much of your AS course was spent studying topics such as biological molecules, cell structure and the processes of diffusion and active transport. These basic processes underpin an understanding of physiological topics. If you consider the example of the digestive system again, it is obvious that an understanding of, for example, the process of digestion requires a knowledge of the structure of molecules such as proteins, lipids and carbohydrates. It also requires an understanding of enzymes and the way in which they work. This basic biology is essential and will come up repeatedly throughout your course. A Level might be based on modules, but this does not mean that once you have studied a subject and taken the relevant unit test you can forget about it. For more about this, see the introduction to Chapter 6, Bringing it all together.

Exercise 5.1 Climbing and diving

The graph in Figure 5.1 shows the oxygen dissociation curve for human haemoglobin.

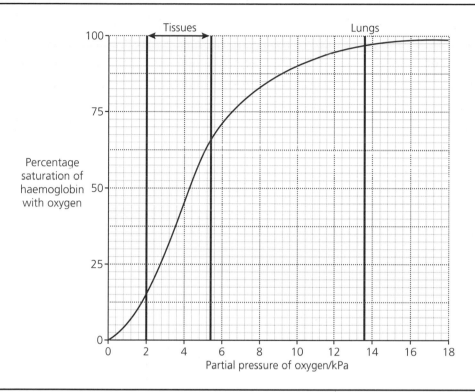

Figure 5.1 *Oxygen dissociation curve for human haemoglobin*

1   (a)   Use Figure 5.1 to explain how haemoglobin transports oxygen from the lungs to the tissues.

(2 marks)

       (b)   Using information from the graph, explain what causes haemoglobin to give up a greater proportion of the oxygen it carries during a period of exercise.

(2 marks)

Air is a mixture of gases. In a mixture of gases, the partial pressure of a particular gas such as oxygen is the proportion of the total pressure that it exerts.

2   Atmospheric pressure decreases with altitude. Giving an explanation for your answer in each case, how would you expect the following to change with an increase in altitude

(a)   the percentage of oxygen in the air

(1 mark)

(b)   the partial pressure of oxygen in the air.

(1 mark)

Table 5.1 shows how the partial pressure of oxygen (pO_2) in the alveoli of the lungs changes with altitude.

Table 5.1 The effect of altitude on the partial pressure of oxygen in the alveoli

Altitude/m	Partial pressure of oxygen/kPa
0 (sea level)	13.8
3100	8.9
5500	6.0
6200	5.3

3   An altitude of 7100 m is thought to be the 'ceiling' for an average person. Survival above this altitude is not possible. Use Figure 5.1 and the data in Table 5.1 to explain why.

(3 marks)

4   People who live at high altitude for any length of time acclimatise to these conditions. The number of red blood cells increases as does the concentration of the substance DPG found inside red blood cells.

(a)   Explain the advantage of having a larger number of red blood cells at high altitudes.

(1 mark)

(b)   High concentrations of DPG result in the oxygen dissociation curve for haemoglobin moving to the right. Explain how this helps a person to live at high altitude.

(3 marks)

Exercise 5.1 *continued*

For a diving animal, problems could potentially result from the increase in the partial pressures of oxygen and nitrogen with depth. Table 5.2 shows how the partial pressures of these gases change with depth.

Table 5.2 The effect of depth on the partial pressures of oxygen and nitrogen

Depth/m	Partial pressure of oxygen/kPa	Partial pressure of nitrogen/kPa
0 (at surface)	21.1	79.8
50	42.4	159.7
100	233.5	878.3
500	1082.6	4072.5

5   Explain why the partial pressure of nitrogen is higher that the partial pressure of oxygen at the surface.

(1 mark)

The diagram in Figure 5.2 shows the relationship between diving depth and nitrogen invasion rate in a human scuba diver and in a whale. Nitrogen invasion rate is the rate at which nitrogen enters the blood from the lungs.

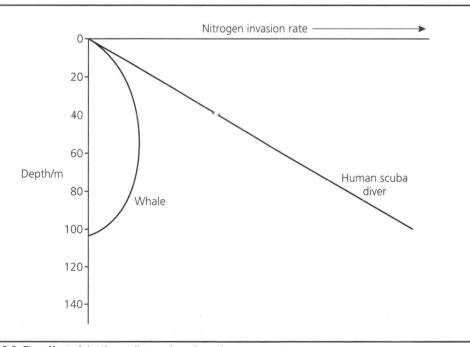

Figure 5.2 The effect of depth on nitrogen invasion rate

Exercise 5.1 *continued*

The lungs of a human scuba diver work normally while diving because the air in the lungs is kept at the pressure of the surroundings by a pressure regulator on the scuba tank.

6 Use the data in Table 5.2 and Figure 5.2 to explain why the nitrogen invasion rate increases with depth in a human scuba diver.

(2 marks)

The nitrogen that enters the blood of a scuba diver dissolves. If the diver returns to the surface too fast, the dissolved nitrogen comes out of solution and forms bubbles in the joints. This causes a very painful condition known as Caisson's disease or 'the bends'. It can be fatal.

7 When a whale dives its lungs collapse completely as the pressure increases with depth.

    (a)   Use this information to explain the shape of the curve for the whale in Figure 5.2.

(2 marks)

    (b)   When a whale surfaces, it does not get the bends. Why not?

(1 mark)

Exercise 5.2 Ventilation in fish

Fish rely on gills for gas exchange. A typical bony fish like a cod or a mackerel has four pairs of gills. The diagram in Figure 5.3 shows how they are arranged. The arrows show the direction of water flow over the gills.

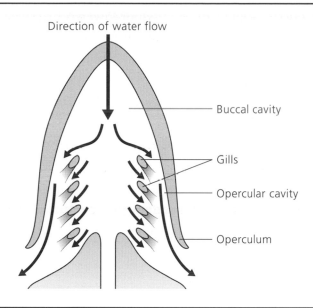

Figure 5.3 *Water flow over the gills of a bony fish*

Fish are able to pump water over the gills and ventilate them. In an investigation, pressure changes in the buccal cavity and the opercular cavity were measured during a number of ventilation cycles. Some of the results of this investigation are shown in Figure 5.4.

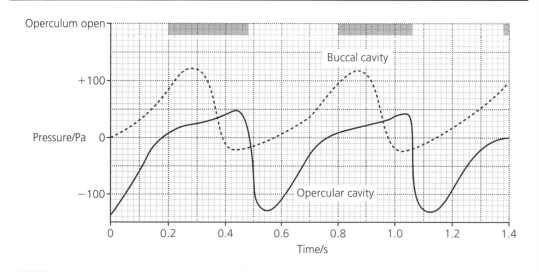

Figure 5.4 *Pressure changes in the buccal and opercular cavities during ventilation cycles in a fish*

Exercise 5.2 *continued*

1 Calculate the rate of ventilation in this fish. Give your answer in cycles per minute.

(1 mark)

2 The floor of the buccal cavity can be raised or lowered by muscles and the mouth can be opened or closed. Use this information together with that in Figure 5.4 to explain what causes water to enter the buccal cavity during ventilation.

(3 marks)

3 During most of the ventilation cycle, water is flowing from the buccal cavity to the opercular cavity over the gills.

   (a)   Give the evidence from Figure 5.4 that would allow you to identify the times when water is flowing from the buccal cavity to the opercular cavity.

(1 mark)

   (b)   Calculate the percentage of a ventilation cycle when water does not flow over the gills. Show your working.

(2 marks)

4 The operculum is able to open and close.

   (a)   Describe the relationship between the pressure in the opercular cavity and the operculum being closed.

(1 mark)

   (b)   Explain why the closing of the operculum is important in maintaining a one-way flow of water over the gills.

(2 marks)

Exercise 5.2 *continued*

5   The sturgeon is a large fish which spends much of its time feeding in the mud at the bottom of lakes and rivers. It takes water in through the operculum before passing it from the opercular cavity to the buccal cavity. Explain how this is an adaptation to the way of life of the sturgeon.

(1 mark)

Mackerel are very active fish. The graph in Figure 5.5 shows the ventilation rate of a mackerel swimming at different speeds.

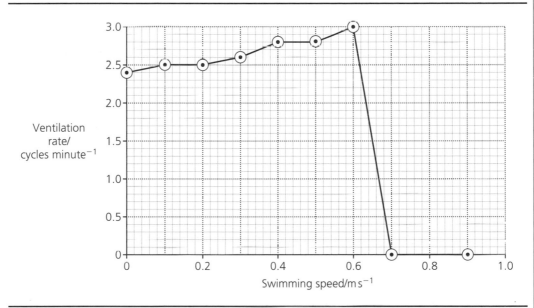

Figure 5.5 *The effect of swimming speed on ventilation rate in a mackerel*

6   (a)   Explain the advantage of the change in ventilation rate as the swimming speed increases from 0 to 0.6 m s^{-1}.

(2 marks)

   (b)   Suggest how the mackerel maintains a flow of water over its gills at swimming speeds over 0.6 m s^{-1}.

(2 marks)

Exercise 5.3 Human milk . . . a perfect food?

In the first few days after the birth of a human baby, the mammary glands secrete a small amount of a thick fluid called colostrum. It contains more protein but less carbohydrate and much less lipid than milk produced later. Colostrum changes to milk three to six days after birth. By the tenth day most of the important changes have taken place and the milk may be described as mature. Table 5.3 shows the amounts of some substances in colostrum and mature milk.

Table 5.3 The composition of human colostrum and mature milk. Amounts are given per 100 cm³

Substance		Colostrum	Mature milk
Proteins	Total protein/g	2.3	0.9
	Casein/mg	140	187
	α immunoglobulin/mg	364	142
	Bile-salt stimulated lipase/mg	n/a	10
Other substances	Urea/mg	10	20
	Amino acids/mg	n/a	52

n/a no data available

1   Early measurements of the total protein content of human milk were based on a method of determining the nitrogen content of milk samples.

   (a)   Explain why nitrogen content can be used as a measure of the protein content of milk.

   (2 marks)

   (b)   The method used estimated the total protein content of human milk as 1.5 g 100 cm⁻³. Explain why this method gave an over-estimate of the total protein content.

   (2 marks)

Exercise 5.3 *continued*

2 Casein is a protein which coagulates when exposed to a low pH and forms curds. These curds are then digested in the baby's gut.

(a) Calculate the percentage of the total protein in mature milk that is casein.

(1 mark)

(b) Cows produce milk which contains a much higher concentration of casein. This results in the formation of larger curds. Explain why the smaller curds produced from human milk are more efficiently digested than the larger curds produced from cow's milk.

(2 marks)

3 Urea has small molecules. It is found in the secretions produced by many of the glands in the body. Suggest an explanation for the presence of urea in human breast milk.

(1 mark)

Exercise 5.4 Diagnosing diabetes

Diabetes mellitus, to give the condition its full name, results from a high blood glucose concentration. This is due either to a lack of the hormone insulin, or to the insulin which the body produces no longer being fully effective. The condition can be treated and diabetics are able to live normal lives, but successful treatment depends on early diagnosis and careful monitoring.

A test frequently used to confirm diabetes is the glucose tolerance test. The person being tested fasts overnight. A sample of blood is taken to determine the fasting blood glucose concentration. He or she is then given a solution containing 75 g of glucose in 300 cm³ of water to drink. Samples of blood are taken every 30 minutes for the next two hours and the blood glucose concentration measured. Some results of glucose tolerance tests are shown in Figure 5.6.

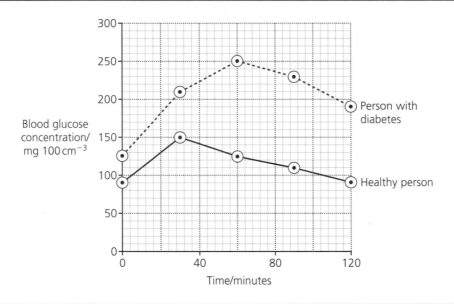

Figure 5.6 *The results of glucose tolerance tests on a healthy person and on a person with moderately severe diabetes*

1   (a)   The person taking the test must remain seated throughout. Describe and explain how physical activity would affect the results of a glucose tolerance test.

(2 marks)

(b)   The curves in Figure 5.6 are for capillary blood. If the blood taken for analysis is removed from a vein, the figures for blood glucose concentration are lower. Explain why.

(2 marks)

Exercise 5.4 *continued*

2 (a) In the healthy person, describe the role of insulin in bringing about the changes in blood glucose concentration shown on the graph.

(2 marks)

(b) Give **two** explanations for the fall in blood glucose concentration after one hour in the person with diabetes.

(2 marks)

3 Suggest why it is recommended that carbohydrates in the diet of a diabetic person should be in the form of starch rather than glucose.

(2 marks)

To be certain that a diabetic person is receiving the correct amount of insulin, it is necessary to monitor the concentration of blood glucose. A disadvantage of monitoring blood glucose concentration is that samples have to be obtained first! It would be more convenient to measure the concentration of glucose in the urine. Figure 5.7 is a graph showing simultaneous measurements of blood glucose concentration and urine test results in a person with diabetes, over a two-week period.

Exercise 5.4 *continued*

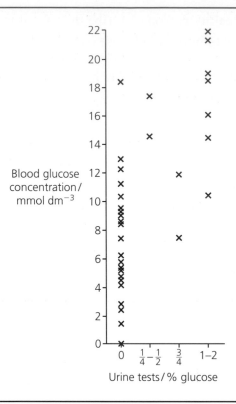

Figure 5.7 *Graph showing the relationship between blood glucose concentration and urine test results for a person with diabetes*

4   (a)   Could urine test results be used as a measure of blood glucose concentration? Explain your answer.

(1 mark)

(b)   Explain how it is possible for this person to have a low blood glucose concentration but a high urine test result.

(2 marks)

5   The renal threshold is the maximum concentration of blood glucose at which the kidney is able to reabsorb all the glucose passing through it. Over the renal threshold, glucose appears in the urine. The renal threshold varies from person to person. Explain how this will affect the usefulness of urine glucose tests.

(2 marks)

Exercise 5.5 The kidney – variations on a theme

Nitrogenous waste is excreted by the kidneys. The main excretory product in a mammal is urea and it has to be removed from the body in solution. The kidneys of mammals are adapted to excreting excess ions and nitrogenous waste while, at the same time, minimising water loss. This ability to produce concentrated urine is due, in part, to the arrangement of the loop of Henle and the collecting ducts.

The maximum concentration of urine that can be produced varies in different species of mammal. For example, many of the small rats and other rodents which live in deserts produce very concentrated urine. The kidneys of these desert-living species have longer loops of Henle than those species which live in conditions where rainfall is higher. The graph in Figure 5.8 shows the relationship between maximum urine concentration and the relative area of the kidney medulla for various species of mammal.

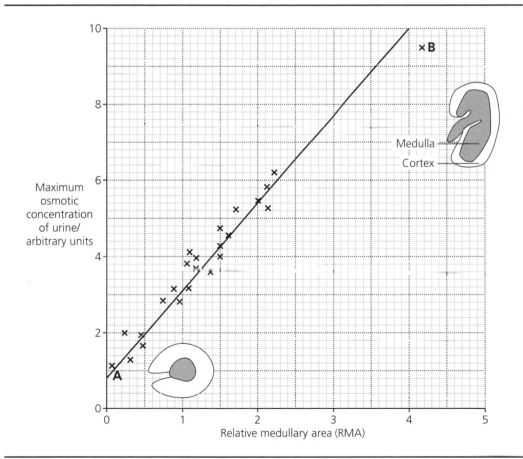

Figure 5.8 *The relationship between the maximum osmotic concentration and the relative medullary area in different species of mammal. The diagrams represent cross-sections through the kidneys of two of the species shown, A and B*

Exercise 5.5 *continued*

1   Write a formula which would allow the relative medullary area (**RMA**) to be calculated. Use A_M to represent the area of the medulla and A_K to represent the area of the whole kidney.

(2 marks)

2   (a)   One of species **A** or **B** lives in desert conditions. The other lives in an area where the rainfall is higher. Which is the desert-living species? Explain your answer in terms of RMA and nephron structure.

(1 mark)

(b)   Explain how a large RMA is related to the ability of the kidney to produce concentrated urine.

(3 marks)

3   Small mammals have cells which contain a greater number of mitochondria relative to their size than larger species. Use this information to explain why small-desert-living species produce more concentrated urine than larger species.

(2 marks)

The common vampire is a small bat found in the tropical areas of Central and South America. It feeds entirely on blood. During the daytime, common vampires roost in caves and hollow trees. After dark, they emerge from their roost and fly off to find a suitable host. They land on it, or nearby, and scuttle across the ground to where it is lying or standing, then use their razor-sharp incisor teeth to cut away a small piece of skin. Blood oozes from the wound and the bat starts to drink this. In about half an hour, it takes in up to 60% of its body mass. Shortly after completing its meal, it flies back to its roost. The graph in Figure 5.9 shows the pattern of urine production by a common vampire bat before and after a meal of blood.

Exercise 5.5 *continued*

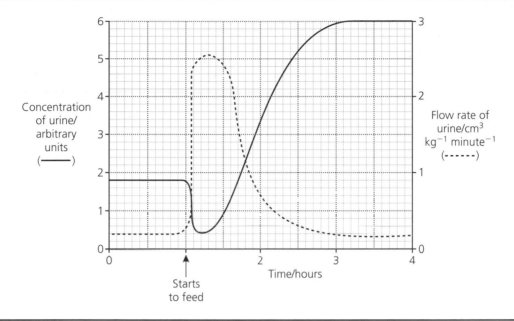

Figure 5.9 *Concentration and flow rate of urine from a vampire bat before, during and after a meal of blood*

4   Describe how the characteristics of vampire bat urine change in the three-hour period after the bat starts to feed.

(2 marks)

5   Use the information in this exercise to explain the advantage to a vampire bat of

   (a)   the change in urine flow rate which takes place immediately after the vampire bat starts feeding

(3 marks)

   (b)   the low urine flow rate for the remainder of the time.

(2 marks)

6   The concentration of urine is linked to the high concentration of protein in a vampire bat's diet.

   (a)   Name **two** proteins which you would expect to find in the bat's food.

(2 marks)

   (b)   The concentration of dissolved substances in the bat's urine changes after feeding. Explain this change in terms of its high protein diet.

(3 marks)

Exercise 5.6 How fast is a nerve impulse?

The speed of a nerve impulse can be estimated in a human by stimulating a nerve at different points and finding how long it takes before electrical events associated with contraction are recorded at the muscle it supplies. Figure 5.10 shows traces obtained from a 20 year-old man. The ulnar nerve was stimulated at the wrist and at the elbow. Electrodes were placed on the skin and used to record the electrical events in the muscle supplied by this nerve. The gap between the stimulus and the events in the muscle is known as the latent period.

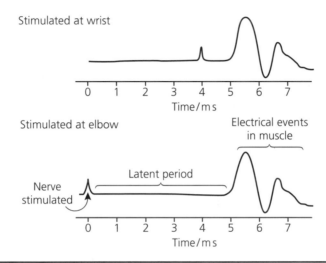

Figure 5.10 *Electrical traces recorded when the ulnar nerve of a 20 year-old man was stimulated at the wrist and at the elbow*

1   (a)   Explain why the latent period was shorter when the nerve was stimulated at the wrist.

(1 mark)

(b)   The distance between the stimulating electrodes at the wrist and at the elbow was 24 cm. Use figure 5.10 to calculate the speed at which the impulse was conducted along the ulnar nerve. Show your working.

(2 marks)

2   (a)   Describe the path of the impulse from the point of stimulation to the muscle.

(1 mark)

(b)   Use your answer to part (a) to explain whether your estimate of conduction speed is accurate.

(1 mark)

Exercise 5.6 *continued*

3 The speed of nerve conduction declines with age. A person aged 60 has a nerve conduction speed about 90% of that of a 20 year-old. Calculate the latent period if the ulnar nerve of a 60 year-old had been stimulated at the elbow. Show your working.

(2 marks)

Figure 5.11 shows how the speed of conduction of a nerve impulse depends on axon diameter.

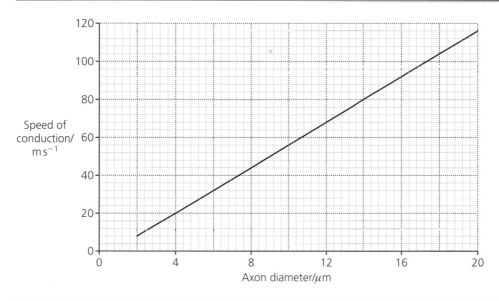

Figure 5.11 *The effect of axon diameter on the speed of conduction of nerve impulses in myelinated nerves from a cat*

4 (a) Describe how a nerve impulse travels along a myelinated axon.

(2 marks)

(b) Studies of neurone structure have shown a direct correlation between diameter of an axon and the length of the Schwann cells surrounding it. Use this information to explain Figure 5.11.

(2 marks)

5 Sloths are mammals which live in trees. They move very slowly. The speed of conduction of nerve impulses in sloth axons is between 6 and 30 m s^{-1}. Does a slow rate of nerve conduction explain the slowness of a sloth's movements? Use information in this exercise to explain your answer.

(2 marks)

Table 5.4 shows the diameter of axons from different organisms and their speed of conduction.

Table 5.4 The effect of axon diameter on speed of conduction in four different species

Species	Axon diameter/μm	Speed of conduction/m s^{-1}
A	9	3
B	10	57
C	40	12
D	90	25

6   Three of the species in Table 5.4 are invertebrates. One is a vertebrate. Which species is the vertebrate? Give a reason for your answer.

(2 marks)

Exercise 5.7 Salt, water and plant growth

The rainfall in many parts of the world is either insufficient or too unreliable to grow crops satisfactorily. One possible solution to this is irrigation. The graph in Figure 5.12 shows wheat yields on plots of land in a semi-arid area in the Middle East. This area receives some rain every year but the pattern of rainfall is very variable and erratic. Some of these plots of land were irrigated and some were not.

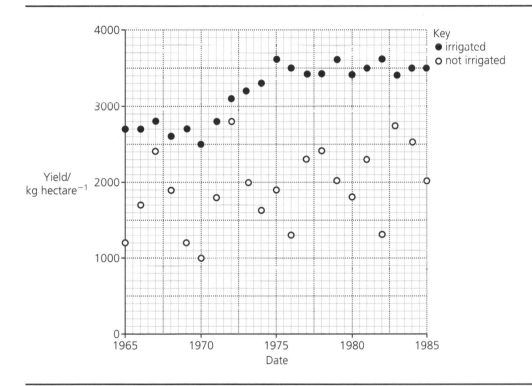

Figure 5.12 *The effect of irrigation on wheat yield*

1 Use Figure 5.12 to give **two** advantages of irrigating wheat.

(2 marks)

2 (a) Sometimes water used for irrigation contains dissolved salts. What will happen to these salts when irrigated soil is exposed to heat from the sun?

(1 mark)

(b) Explain how a high salt concentration in the soil will affect the amount of water that a plant can extract from the soil.

(2 marks)

Exercise 5.7 *continued*

Wheat plants were grown in solutions with different salt concentrations. At each concentration, the rate of carbon dioxide uptake and the water potential of the leaf cells were measured. The results are shown in Table 5.5.

Table 5.5 The effect of salt concentration on rate of carbon dioxide uptake and the water potential of leaf cells in wheat plants

Salt concentration/mmol dm^{-3}	Rate of carbon dioxide uptake/mg dm^{-2} h^{-1}	Water potential of leaf cells/MPa
20	21.4	−0.36
100	21.5	−0.39
180	22.4	−0.89
260	24.7	−1.31
340	29.4	−1.42

3 Describe and suggest an explanation for the relationship between salt concentration and the water potential of the leaf cells.

(3 marks)

4 (a) What, precisely, does rate of carbon dioxide uptake measure?

(1 mark)

(b) The wheat plants growing in the more concentrated salt solutions had a lower rate of growth than those growing in the less concentrated solutions. Do the data in the table support the view that this lower rate of growth is due to a lower rate of photosynthesis? Explain your answer.

(1 mark)

Salt marshes are found in coastal areas. They are frequently flooded with sea water. Some plants, such as sea plantain, are adapted to living in these conditions. In an investigation, five-week old sea plantain seedlings were grown in aerated solutions containing different concentrations of sodium chloride. After seven days the plants were analysed for water, sodium ions and sorbitol. Sorbitol is a soluble organic substance. The results are shown in Figure 5.13.

Exercise 5.7 *continued*

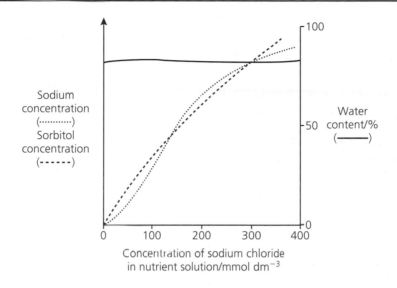

Figure 5.13 *The effect of the sodium concentration on water content, and sodium and sorbitol concentration in seedlings of sea plantain*

5 Explain why the solution in which the sea plantain seedlings were growing was aerated.

(1 mark)

6 (a) Use evidence from Figure 5.13 to explain the effect of increasing salt concentration on the water content of the sea plantain seedlings.

(2 marks)

(b) Suggest how the responses to salinity shown in Figure 5.13 may enable sea plantain to survive in a salt marsh without wilting.

(2 marks)

Exercise 5.8 The production of sperm cells

One of the important ways in which the testes of a mammal differ from the ovaries is in their ability to produce huge numbers of sex cells throughout the reproductive life of the organism. In a human male, for example, over a thousand sperm cells are produced every second. Sperm production involves two stages. In the first, mitosis takes place and the cells lining the seminiferous tubules multiply. Meiosis then results in some of these cells developing into sperms. Nuclei from cells undergoing meiosis were examined and the amount of DNA they contained was estimated. The method used involved staining the DNA and measuring the amount of light it absorbed. Figure 5.14 shows some results from this investigation.

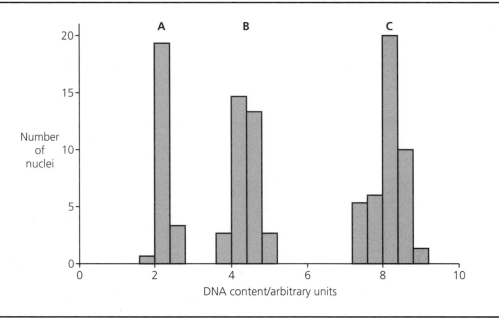

Figure 5.14 *The amount of DNA in individual nuclei from a mouse testis*

1   (a)   Which group of cells, **A**, **B** or **C**, represents cells before they begin to divide? Give a reason for your answer.

(1 mark)

(b)   Use your knowledge of chromosome behaviour in meiosis to account for the amounts of DNA in the other two groups of cells.

(2 marks)

2   The measurements for each group of cells show variation. Suggest **two** explanations for this variation.

(2 marks)

Exercise 5.8 *continued*

3   Describe the pattern of DNA content you would expect for cells undergoing mitosis. Explain your answer.

(2 marks)

The process of sperm formation is controlled by hormones. Table 5.6 shows some of the hormones produced by a male mammal and their role in sperm production.

Table 5.6 The role of some hormones in sperm production

Hormone	Role in sperm production
FSH	Stimulates production of sperm cells
LH	Stimulates production of testosterone by Leydig cells in the testis
Testosterone	Stimulates production of sperm cells

The graph in Figure 5.15 shows fluctuations in the concentration of LH and testosterone in the blood of a male sheep.

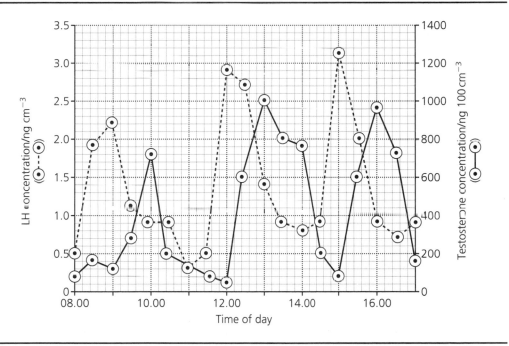

Figure 5.15 *Fluctuations in the concentration of LH and testosterone in blood from the jugular vein of a male sheep*

Exercise 5.8 *continued*

4   (a)   Use information from Table 5.6 to explain the pattern of
             testosterone secretion shown in Figure 5.15.

                                                                   (2 marks)

     (b)   Some men are infertile because they do not produce sufficient LH.
             They can be treated with extra LH. This treatment produces an
             increase in the amount of smooth endoplasmic reticulum and the
             number of mitochondria in the Leydig cells. Suggest how this
             might lead to an increase in testosterone secretion by these cells.

                                                                   (4 marks)

5 Explain how Figure 5.15 provides evidence that negative feedback controls
   blood LH concentration.

                                                                   (2 marks)

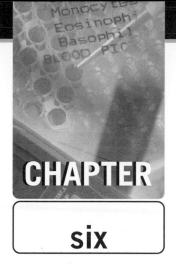

Bringing it all together

One of the most important features of an A Level course is that you will be required to bring together knowledge from different areas of biology. We call questions which ask you to do this 'synoptic' questions. They are very important because they account for 40% of all the marks available for the A2 part of your A Level. The exercises in this chapter are all synoptic because they all require you to bring together information from different parts of the specification. If you want to succeed, you should really start to think about this aspect of your biology as early as possible. It is far too important to leave to the end so, if you have just started your A2 work, what can you usefully do?

Themes and concepts

When we look at any of the body systems in detail we find that we are continually drawing on knowledge acquired earlier in the course to explain different aspects of the way in which they function. The method by which the nervous system transmits impulses, for example, depends on the passage of ions through the plasma membranes of nerve cells. This involves basic processes such as diffusion and active transport. In a similar way, digestion is based on the hydrolysis of large biological molecules such as proteins, carbohydrates and lipids. Look at Box 8 on page 116. This gives you a list of basic principles and concepts. All of them are ideas which come up time and time again on an A Level course and we know that, whatever physiological topic we are studying, we are going to require knowledge of some of these to gain a complete understanding of the underlying biology.

Concept plans

It is obviously very important to be as familiar as possible with the links between different aspects of biology. As we said at the start of this chapter, you need to start thinking about these links as early as possible, preferably right at the beginning of your A2 year. One thing that might help you is to produce a concept plan at the end of each topic you have studied. This should help to show these links. We have called it a concept plan here, but you could also refer to it as a mind map or, for that matter, a spider diagram. It makes no difference. We will look at how we might set about constructing such a plan for the digestive system. The completed plan is shown in Figure 6.1. You will need to look at this as you read through the bullet points below.

- Start by summarising the content of the topic you have just studied with a few simple, clear statements. You may find your specification helpful here. Write these statements on the left-hand side of a sheet of paper and space them out well.

- The next step is to produce as many links as you can between your summary statements and the principles and concepts listed in Box 8. Concentrate on establishing the obvious links. You will not be able to connect all the items in Box 8 with each summary statement, but that does not matter. These links are shown in Figure 6.1 in the second column.

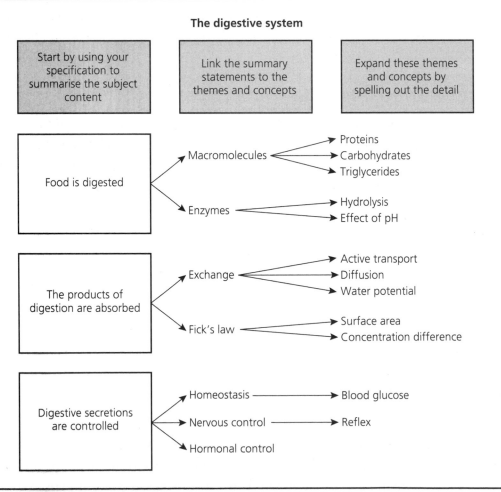

Figure 6.1 *Constructing a concept plan*

- Finally, expand the topics you have transferred from Box 8 and spell out the precise aspects that are important. This has been done in the third column of Figure 6.1.

So . . . you have your concept plan or whatever you wish to call it. Put it in the front of the relevant section of your notes. You should find it very useful when you come to revise. The first column should come in useful in helping to make sure that you have a sound knowledge and understanding of the topic you have studied. The rest of the sheet will act as a guide to the other areas of your specification that you ought to look at. It might have been a long time since you last looked at them!

Before you start on the exercises in this chapter, here are a few hints which might help you to avoid some of the problems which arise most frequently when answering synoptic questions.

- Think about synoptic questions at the start of your course . . . it is far too important to leave until the end.

- Expect the unfamiliar. Look through the subjects of these exercises: parasitic worms, butterflies . . . They are probably not on your specification. It doesn't matter. The underlying biology is. Remember, if you haven't encountered this particular bit of biology before, it is quite likely that no one else has either.

- Do not invent silly biology! You should have encountered most of what you need to answer the questions in these exercises before. The key question that should always be going through your mind is 'Which bit of biology, that I have studied before, do I want here?' Sort that out and you are halfway there.

- You still need all those data handling skills. You are going to have to describe patterns and trends in data, explain and interpret graphs and tables, and evaluate or judge the worth of information. Make sure you can.

BOX 8 Principles and concepts

You will encounter the ideas in this box frequently. In order to understand the processes which take place in the systems which make up a living organism, you will often need to make use of them.

- Cell **organelles** and their functions
- The **exchange** of substances across cell membranes
- The ratio of **surface area to volume**
- **Fick's law** – the factors which affect the rate of diffusion
- The shape, structure and function of **protein molecules**
- **Macromolecules**, condensation and hydrolysis
- **Enzymes** and the factors which affect them
- Transfer of **energy**
- The relationship between **structure and function**
- **Homeostasis** and negative feedback
- Nervous and chemical **control**
- **Selection** and evolution

This list is not meant to be complete. See if you can add to it.

You should use the ideas in this box when you build up concept plans such as that shown in Figure 6.1. The key words have been emboldened. If you use just these words in the first column of your concept plan it will allow you a little more room.

Exercise 6.1 Living in the gut

A parasite is an organism which lives inside or on the surface of another organism and often causes harm. There are many different types of parasite. Humans, for example, may have fleas and lice in their hair, malarial parasites in the blood and a variety of worms and flukes in the gut. It might be thought that parasitism is an ideal way of life, living in a regulated environment with a plentiful supply of food. There are, however, problems associated with a parasitic existence. A parasite has to infect not only the right host but it must also reach the right part of that host. It also has to be adapted to living within what can be a very demanding environment.

Cyathocotyle bushiensis is a fluke which lives in the alimentary canal of its mammalian host. It enters as a cyst and then hatches to produce the fluke. The number of flukes hatching after being treated in different ways was investigated. Some results from this investigation are shown in Figure 6.2.

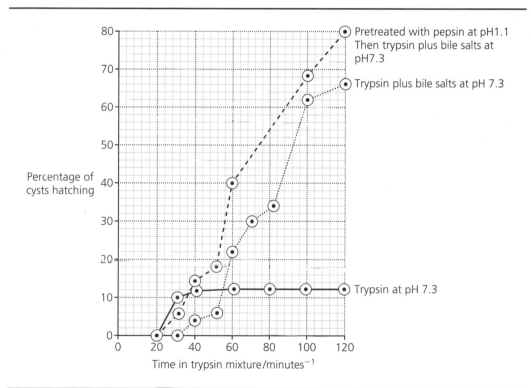

Figure 6.2 *Graph showing the effects of different treatments on the percentage of cysts of* Cyathocotyle bushiensis *which hatched*

1   Describe how you could demonstrate that it was the enzymes, rather than the accompanying pH changes, which stimulated hatching.

(1 mark)

Exercise 6.1 *continued*

2 Describe how pre-treating with pepsin affects hatching when the cysts are incubated with trypsin and bile.

(2 marks)

3 (a) Use the graph to predict in which part of the alimentary canal you would expect most cysts to hatch. Explain the evidence for your answer.

(2 marks)

(b) Suggest why few cysts would hatch when swallowed by an invertebrate.

(2 marks)

Until recently it was thought that conditions in the lumen of the small intestine of a mammal were anaerobic, with oxygen only present near the gut wall.

4 Explain why oxygen would be present near the gut wall.

(2 marks)

More recent work, has shown that the concentration of oxygen in the intestinal lumen is similar to that in venous blood. There is also a high concentration of carbon dioxide present.

5 Suggest **two** explanations for the high concentration of carbon dioxide in the lumen of the intestine.

(2 marks)

Exercise 6.1 *continued*

Some of the parasitic worms which live in the intestine have a form of haemoglobin. The graph in Figure 6.3 shows the oxygen dissociation curves for this haemoglobin at high and low partial pressures of carbon dioxide.

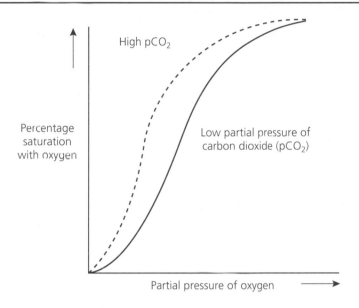

Figure 6.3 *Oxygen dissociation curves for haemoglobin from a parasitic worm*

6   (a)   The partial pressure of oxygen at which haemoglobin is 50% saturated with oxygen is called the P_{50}. How does a high partial pressure of carbon dioxide affect the P_{50} of the parasite haemoglobin?

(1 mark)

       (b)   Explain how this effect could be an advantage to the parasite.

(3 marks)

Exercise 6.2 Growth and development in insects

Insects such as butterflies and moths have complex life-cycles. The adult female lays eggs which hatch to produce larvae. These larvae feed and grow, eventually moulting into pupae. Inside the pupa, larval organs are broken down and adult organs are formed. Eventually the pupal case splits and a new adult emerges.

In butterflies, the larva or caterpillar spends much of its time eating. One way of measuring the food eaten is to calculate the consumption index (C.I.). It is calculated from the formula

$$C.I. = \frac{\text{mass of food eaten}}{\text{mean mass of insect during feeding period} \times \text{length of feeding period (in days)}}$$

1 Suggest an advantage of measuring the food eaten as the consumption index rather than just the mass of food.

(1 mark)

2 An insect larva eats twice its body mass in food in a day. Calculate its consumption index. Explain how you worked out your answer.

(2 marks)

The food eaten by an insect larva may differ from that eaten by the same insect when it is an adult. Table 6.1 shows the enzymes in the midgut of a species of butterfly as a larva and as an adult.

Table 6.1 Digestive enzymes present in the midgut of a species of butterfly

Stage	Food	Enzyme				
		Protease	**Lipase**	**Amylase**	**Sucrase**	**Maltase**
Larva	Leaves	✔	✔	✔	✔	✔
Adult	Nectar	✗	✗	✗	✔	✗

✔ Enzyme present
✗ Enzyme absent

3 What carbohydrate, present in large amounts in leaves, cannot be digested by the enzymes produced by the larva of this species of butterfly?

(1 mark)

Exercise 6.2 *continued*

4 The enzymes in the midgut are linked to the nutrients present in the insect's food. Describe, as precisely as possible, where you would expect to find the following in leaf tissue

(a) lipids

(2 marks)

(b) starch.

(2 marks)

5 A butterfly larva usually stays on its food plant. It is the growth stage of the insect. The adult does not grow. However, it is able to fly and is thus able to find new host plants on which to lay its eggs. Explain how the enzymes present in the midgut and the food eaten are linked to

(a) growth in a butterfly larva

(3 marks)

(b) flight in an adult butterfly.

(2 marks)

Exercise 6.2 *continued*

Figure 6.4 shows some of the changes which take place inside the cells of an insect pupa.

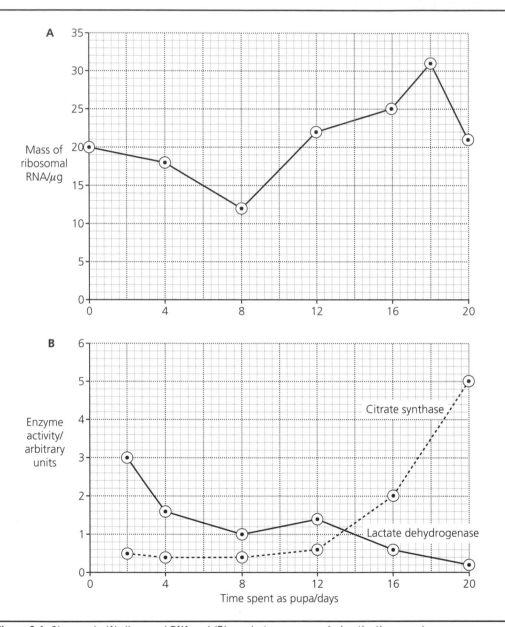

Figure 6.4 *Changes in (A) ribosomal RNA and (B) respiratory enzymes during the time spent as a pupa*

6   Explain the link between the change in mass of ribosomal RNA between days 8 and 18 and the formation of adult organs.

(2 marks)

Exercise 6.2 *continued*

7   The enzymes citrate synthase and lactate dehydrogenase are involved in
different stages of respiration. Citrate synthase catalyses the reaction in which
acetylcoenzyme A combines with the four-carbon compound, oxaloacetate, to
produce citrate, a compound which contains six carbon atoms. Lactate
dehydrogenase catalyses the conversion of pyruvate to lactate.

   (a)   Use this information to suggest and explain how oxygen
consumption per gram of glucose oxidised changes during the time
the insect spends as a pupa.

(2 marks)

   (b)   The increase in citrate synthase activity has been linked to the
formation of tracheae in the pupa. Explain this link.

(2 marks)

Exercise 6.3 Living in a pocket

Kangaroos are marsupials. Marsupials differ from other mammals in their method of reproduction. A young kangaroo weighs under a gram at birth and is very immature. With help from its mother, it moves up into her pouch where it immediately attaches itself to a nipple. Within the pouch, it continues to grow and develop, getting all the nutrients it requires from its mother's milk. After some 250 days, when it weighs over 6 kg, it leaves the pouch. Some of the events associated with the time a young kangaroo spends in the pouch are shown in Figure 6.5.

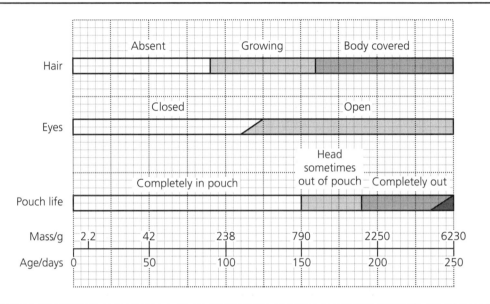

Figure 6.5 *Some of the events associated with the development of a young red kangaroo in the pouch of its mother*

During the time that the young kangaroo spends in its mother's pouch, the composition of her milk changes considerably. Some of these changes are shown in the graphs in Figure 6.6.

Exercise 6.3 *continued*

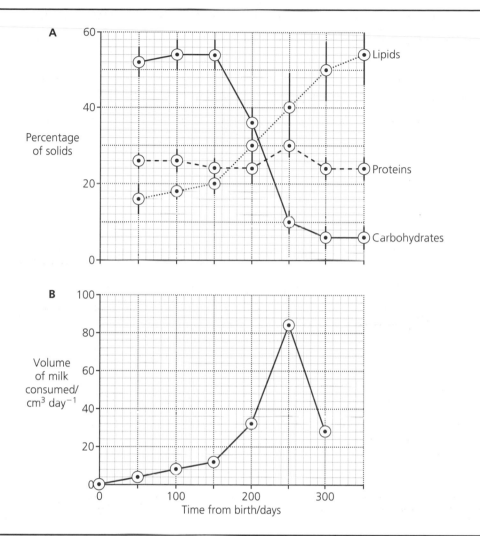

Figure 6.6 *Changes in (A) the composition of kangaroo milk and (B) the volume milk consumed by a young kangaroo*

1   The bars on Figure 6.6 (A) represent standard deviation. What can you conclude from the length of these bars about the composition of the milk?

(1 mark)

2   (a)   Describe how the daily protein intake of the young kangaroo changes over the period shown in Figure 6.6.

(2 marks)

(b)   Describe and explain how the daily protein intake of the young kangaroo is related to its rate of growth.

(2 marks)

Exercise 6.3 *continued*

3   There is a large increase in the concentration of sulphur-containing amino acids in the milk of a kangaroo at around 80 days. Use Figure 6.5 to suggest the advantage of this to the young kangaroo.

(2 marks)

4   Describe and explain how you would expect the respiratory quotient (RQ) of the young kangaroo to change over the period shown in Figure 6.6.

(3 marks)

Table 6.2 shows the partial pressures of some gases in the atmospheric air and in the air inside a kangaroo pouch. All readings were taken at sea level.

Table 6.2 The partial pressure of oxygen and carbon dioxide in atmospheric air and in the air inside a kangaroo pouch

Source of air sample	Partial pressure/kPa	
	Oxygen	Carbon dioxide
Atmosphere	21.1	0.03
Inside kangaroo pouch	15.8	5.32

5   (a)   Explain the difference between the partial pressures of these gases in atmospheric air and in the air inside a kangaroo pouch.

(3 marks)

(b)   Suggest the advantage to a young kangaroo of having a type of haemoglobin which has an oxygen dissociation curve to the left of that for the haemoglobin of an adult.

(2 marks)

Exercise 6.4 G6PD

Glucose 6-phosphate dehydrogenase (G6PD) is an enzyme. Figure 6.7 shows part of the metabolic pathway in which it is involved.

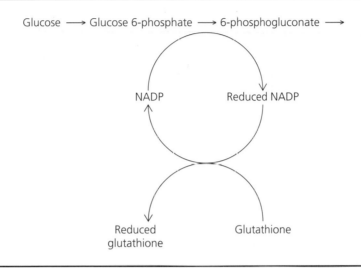

Glucose ⟶ Glucose 6-phosphate ⟶ 6-phosphogluconate ⟶

NADP   Reduced NADP

Reduced glutathione   Glutathione

Figure 6.7 *Part of the metabolic pathway in which the enzyme G6PD is involved*

1 Explain why this enzyme is called glucose 6-phosphate dehydrogenase.

(2 marks)

This metabolic pathway is found in red blood cells. The reduced glutathione which is produced helps to protect the cell against oxidising substances, some of which are produced in the cell as metabolites. These oxidising substances would otherwise lead to the breakdown of red blood cells. The gene which codes for G6PD is found on the X-chromosome. Table 6.3 shows some data about the DNA coding for this enzyme and the mRNA which is transcribed from it.

Table 6.3 Some data concerning the gene coding for G6PD and the mRNA transcribed from it

DNA	Length of gene (number of nucleotides on sense strand)	18500
	Number of exons	13
	Number of introns	12
mRNA	Length of molecule (number of nucleotides)	2268

2 (a) Use the data in Table 6.3 to estimate the number of amino acids in a G6PD molecule. Explain how you arrived at your answer.

(2 marks)

(b)   Explain the difference between the length of the gene and the
        length of the mRNA molecule which is transcribed from this gene.

(1 mark)

The drug primaquine was introduced in 1925 to combat malaria. It is an
oxidising substance. Some people, referred to as being primaquine sensitive,
developed haemolytic anaemia as a result of taking this drug. Haemolytic
anaemia is caused by the breakdown of red blood cells. Two or three days after
starting to take primaquine these people developed a range of symptoms including
the production of dark-coloured urine.

3   Suggest why people who developed haemolytic anaemia as a result of taking
    primaquine produced dark-coloured urine.

(2 marks)

4   Experiments involving labelling red blood cells with a radioactive isotope of
    iron showed that the cells in a sensitive person were broken down when
    63 – 66 days old but not when 3 – 21 days old. Use this information to suggest
    why primaquine sensitive people who developed haemolytic anaemia gradually
    recovered even when drug treatment was continued.

(2 marks)

These early observations led to the discovery that people who were primaquine
sensitive possessed a less active form of G6PD. More evidence from this came
from experiments involving incubation of sensitive and non-sensitive red blood
cells with and without primaquine in a solution containing glucose. The graph in
Figure 6.8 shows some results from this investigation.

5   Explain how a mutation could result in a less active form of the enzyme
    G6PD.

(4 marks)

6   Give the evidence from Figure 6.8 that G6PD is not as effective in sensitive
    cells. Use the metabolic pathway in Figure 6.7 to help you to explain your
    answer.

(2 marks)

Exercise 6.4 *continued*

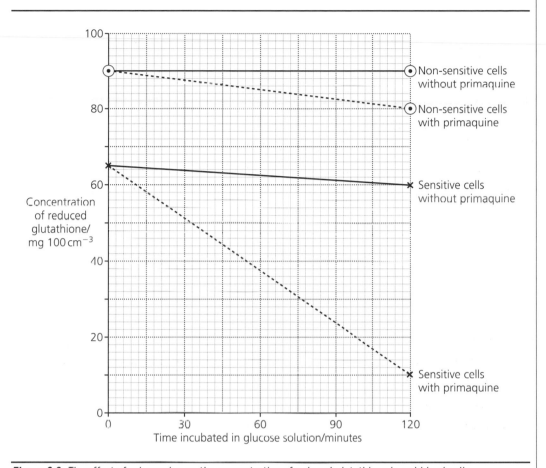

Figure 6.8 *The effect of primaquine on the concentration of reduced glutathione in red blood cells*

7 (a)   When no glucose was present, the concentration of reduced glutathione in non-sensitive cells fell when the cells were incubated with primaquine. Explain why.

(2 marks)

(b)   Explain why the concentration of reduced glutathione does not fall in non-sensitive cells incubated with primaquine when glucose is present.

(1 mark)

8 Explain the results shown on the graph for sensitive cells incubated with primaquine.

(2 marks)

Exercise 6.5 Earthworms and water balance

Up to 85% of the body mass of an earthworm is water. In a dry environment an earthworm rapidly loses water through its body surface, in urine and in faeces. The excretory organs of earthworms are nephridia. Each nephridium consists of a long tube divided into several different regions. Fluid is filtered from the body cavity into one end of this tube. It passes along the nephridium and out of the earthworm through a small pore. The graph in Figure 6.9 shows the concentration of the fluid at various points along a nephridium.

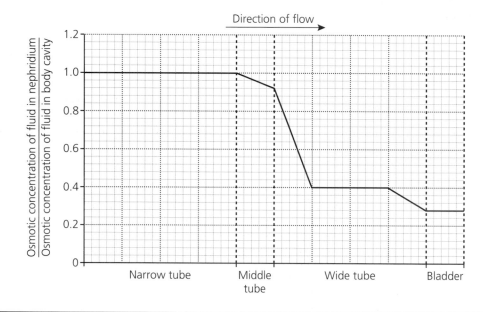

Figure 6.9 *The concentration of fluid at various points along a nephridium of an earthworm*

1   In what part of the nephridium is the concentration of the fluid the same as the concentration of the fluid in the body cavity?

(1 mark)

2   (a)   Describe how the water potential of the fluid in the nephridium changes as it passes along the wide tube.

(2 marks)

        (b)   Explain how active uptake of sodium ions may account for the change in the concentration of the fluid in the wide tube. In your answer, you should also refer to chloride ions and water.

(3 marks)

Exercise 6.5 *continued*

3   An analysis of earthworm urine showed that it contained between 31.7 and 39.1 mg of urea per 100 cm³. Explain what causes the actual amount to vary with temperature.

(2 marks)

4   Some tropical earthworms live in very dry conditions. These species have nephridia which do not open on the outside of the worm. Instead, they open into the gut. Suggest how this is an advantage to worms living in very dry conditions.

(2 marks)

The graphs in Figure 6.10 show the distribution of a species of earthworm at different depths in the soil together with temperature and soil moisture content.

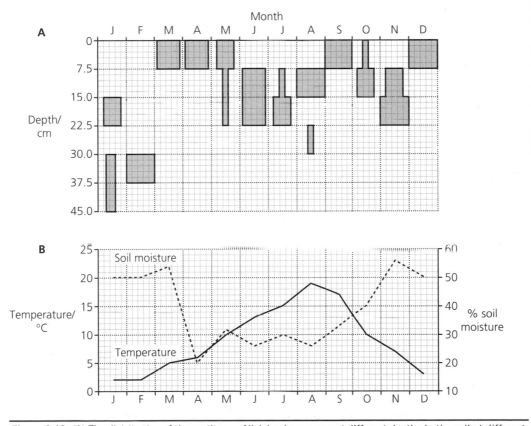

Figure 6.10 *(A) The distribution of the earthworm* Allolobophora rosea *at different depths in the soil at different times of the year. Graph (B) shows soil moisture and temperature at a depth of 10 cm*

Exercise 6.5 *continued*

5   The moisture content of a soil sample can be determined by gently heating the sample. To ensure that results are representative and reliable, samples should be collected at random and care taken to ensure that all the moisture is removed when the sample is heated. Explain how you would ensure that:

   (a)   the samples were collected at random

(3 marks)

   (b)   all the water in a sample was removed.

(1 mark)

6   (a)   Explain the advantage to this species of earthworm of its distribution in the soil during the summer months of June, July and August.

(2 marks)

   (b)   During the summer months, earthworms may aestivate. They curl their bodies into a ball and their respiration rate falls. Explain how aestivation helps earthworms to reduce water loss.

(4 marks)

Exercise 6.6 Plant suckers

Aphids are tiny insects which feed on plants. They have piercing mouthparts which they insert into the phloem of a food plant before sucking up the sap it contains. One species of aphid is the sycamore aphid. As its name suggests, it feeds on sycamore trees. Figure 6.11 shows the mass and reproductive rates of sycamore aphids at different times of the year. It also shows changes in the concentration of nitrogen-containing compounds in the phloem.

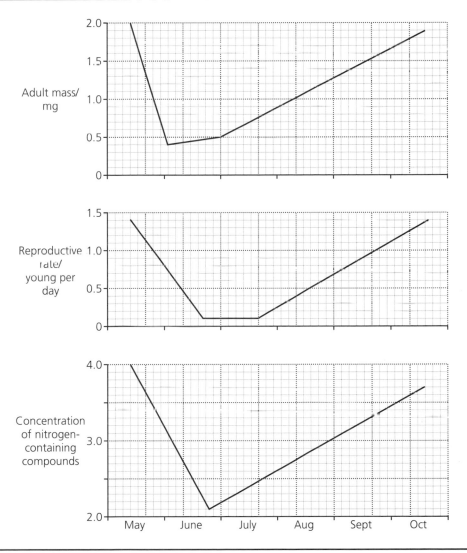

Figure 6.11 *The mass and reproductive rate of sycamore aphids and the concentration of nitrogen-containing compounds in the phloem of sycamore leave at different times of the year*

Exercise 6.6 *continued*

1   Name **one** nitrogen-containing compound that you would expect to find in the phloem.

(1 mark)

2   Explain why the concentration of nitrogen-containing compounds in the phloem is high

   (a)   in early May when buds are bursting and leaves starting to grow

(2 marks)

   (b)   in October when leaves are changing colour.

(2 marks)

3   Describe and explain the relationship between the concentration of nitrogen-containing compounds in the phloem and the reproductive rate of the aphids.

(3 marks)

4   Some species of aphid move from trees to small herbaceous plants in the summer. One of these is the cherry-oat aphid. It spends spring and autumn feeding on cherry trees and the summer on oats. Suggest the advantage to the aphid of changing its host plant in this way.

(1 mark)

Small wasps parasitise aphids. The female wasp lays an egg inside its aphid host. The egg hatches and the resulting wasp larva feeds on the internal organs of the aphid, eventually killing it. Before they can lay eggs, wasps must locate their hosts. Table 6.4 shows the results of some choice-chamber experiments in which the wasp parasites were offered a choice between an experimental chamber and a control chamber.

Exercise 6.6 *continued*

Table 6.4 The results of choice-chamber experiments with a species of wasp which parasitises aphids

Material in experimental chamber	Sex of wasp	Number of wasps in	
		Experimental chamber	Control chamber
Cabbage leaves	Female	84	6
	Male	48	42
Potato leaves	Female	38	41
	Male	42	38
Aphids on damp filter paper	Female	87	8
	Male	41	37
Live female wasps	Female	38	41
	Male	52	16

5 (a) Suggest a suitable control for these experiments.

(1 mark)

(b) Explain why a statistical test should be used to analyse the results shown in the table.

(2 marks)

6 (a) Use information from Table 6.4 to suggest how male wasps locate females.

(1 mark)

(b) In a separate investigation, the percentage of aphids parasitised by this wasp was determined in two adjacent fields. In one field, cabbages were growing; in the other, the crop was potatoes. The results of this investigation are shown in Figure 6.12.

Exercise 6.6 *continued*

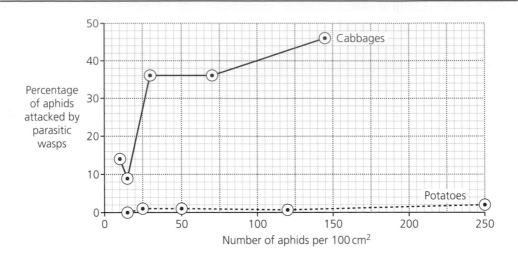

Figure 6.12 *The effect of aphid population density on the percentage attacked by parasitic wasps in adjacent fields containing cabbages and potatoes*

7 Use information from Table 6.4 to suggest a suitable explanation for the difference in the percentage of aphids parasitised in the two fields.

(2 marks)

Exercise 6.7 Carbohydrates and exercise

Running a marathon requires a lot of work. The energy for this work comes from respiration which, in endurance events like marathons, largely depends on carbohydrates. Sports physiologists have investigated the link between carbohydrate in the diet and performance.

In one investigation a group of six athletes ate a mixed diet for a period of three days and then cycled on exercise bicycles until they were exhausted. The same athletes then ate a diet low in carbohydrate but high in protein and lipid for the next three days before again cycling to exhaustion. The procedure was repeated a third time with a high carbohydrate diet. Each time muscle samples were removed immediately before and after exercise and the amount of glycogen present was determined. The results of the investigation are shown in Figure 6.13.

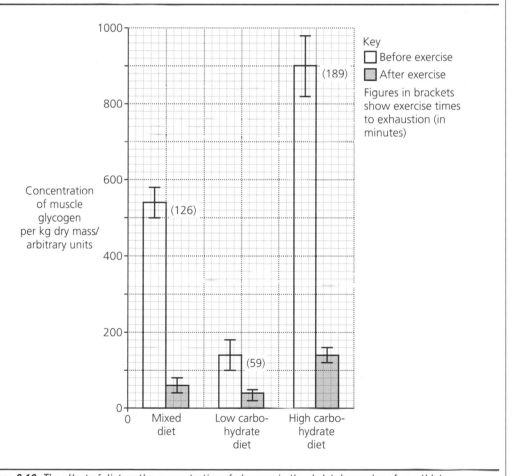

Figure 6.13 *The effect of diet on the concentration of glycogen in the skeletal muscles of an athlete*

1 Explain the advantage in this investigation of

    (a)   using the same group of athletes each time

(2 marks)

    (b)   giving the amount of glycogen in the muscle *per kilogram dry mass*.
(2 marks)

2 The data used to draw the columns on the barchart are the mean values for the group. Clearly, if we had selected a different group of athletes, the mean values would have been different. The standard error is a way of describing the difference in the means we might expect if we did this. The bars on the columns represent the standard error. Use this information to suggest conclusions that can be drawn about the difference in glycogen concentration before and after exercise in athletes on a low carbohydrate diet.

(2 marks)

3 Did a high carbohydrate diet improve the performance of these athletes? Explain your answer and use suitable calculations from Figure 6.13 to support it.
(4 marks)

There are a number of factors that contribute to fatigue during strenuous exercise. A lack of muscle glycogen is one. Another is the dehydration which results from increased sweating. Sports drinks which contain glucose and sodium and potassium salts provide an athlete with both fuel and fluid, and ought to delay the onset of fatigue.

4 Sports drinks must be properly formulated. If they contain too much glucose, the rate at which fluid is absorbed from the small intestine into the blood slows.

    (a)   Describe the part played by osmosis in the absorption of water from the small intestine.

(2 marks)

    (b)   Suggest why too much glucose in a sports drink will slow the rate at which fluid is absorbed from the small intestine.

(2 marks)

Exercise 6.7 *continued*

In an investigation into the effect of sports drinks during prolonged exercise, two groups of athletes were allowed to run on a treadmill until they were exhausted. Members of the first group were allowed to drink sports drinks while they were running. Those in the second group were only allowed to drink water. Table 6.5 shows the concentration of glycogen in muscle samples taken from these athletes before and after the period of exercise.

Table 6.5 The effect of sports' drinks on the glycogen concentration of muscle

Drink provided	Mean glycogen concentration per kg dry mass/arbitrary units		
	at start	after 104 minutes	after 132 minutes
Sports drink	445	138	116
Water	442	114	*

*Note   After 104 minutes the experiment was discontinued for this group. The athletes were too fatigued to continue.

5 Explain the difference in the data for glycogen concentration for the two groups at the start of the investigation.

(1 mark)

6 What can you conclude from the results of this investigation about the effect of the sports drink on endurance? Suggest an explanation for your answer.

(2 marks)

Exercise 6.8 Suicide bags

Lysosomes are cell organelles which contain a number of digestive enzymes. They are difficult to identify because they have very few distinctive features. However, they show a positive reaction to the test for the enzyme acid phosphatase. This test is often used to show the presence of lysosomes in sections of tissue viewed with an electron microscope. A section is incubated with a solution of lead acetate. This solution contains an appropriate substrate for the enzyme and is kept at a pH of 5.5. Phosphate ions are released from the substrate and these combine with lead acetate to form lead phosphate. The section is then prepared for examination with an electron microscope.

1 The solution added to the tissue is kept at pH 5.5. Explain what would happen to the rate of reaction of the enzyme if the pH were to differ from this value.

(3 marks)

2 Explain how this technique would enable a biologist to locate lysosomes on an electron micrograph.

(2 marks)

Lysosomes have a number of functions. They are involved in intracellular digestion of food and microorganisms. They remove faulty proteins from cells and they also break down cell components which are no longer needed – this is why they are sometimes called 'suicide bags'. In all of these processes, macromolecules are broken down by digestive enzymes in the lysosomes to produce smaller molecules. These smaller molecules pass out through the lysosome membrane into the cytoplasm.

Young frogs are tadpoles. When tadpoles turn into frogs, their tails become shorter and finally disappear. Figure 6.14 shows how the concentration of a protein-digesting enzyme found in lysosomes changes as a tadpole matures into a frog.

Exercise 6.8 *continued*

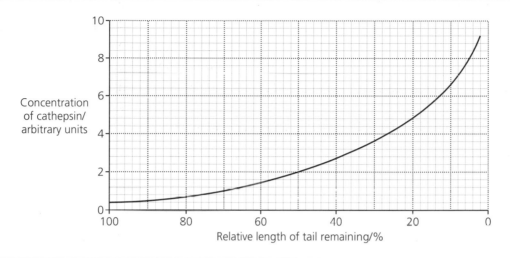

Figure 6.14 *Change in the concentration of a cathepsin, a lysosomal protein-digesting enzyme, with decreasing tail length in developing frogs*

3   One explanation for the increase in cathepsin concentration in Figure 6.14 is that more enzyme is being produced. Suggest another explanation.

(2 marks)

4   (a)   Explain the advantage to the developing frog of the proteins in the tadpole's tail breaking down and the products being absorbed.

(2 marks)

   (b)   Apart from protein-digesting enzymes, suggest one other type of digestive enzyme that you would expect to show a similar change in concentration as the tadpole's tail shortened. Give a reason for your answer.

(2 marks)

Some features of an organism's biology are very sensitive to the effects of pollutants such as metal ions. This means that these features can sometimes be used to detect very low concentrations of these substances in the environment. The effects of low concentrations of mercury ions (Hg^{2+}) and copper ions (Cu^{2+}) in seawater were investigated in a small marine animal, *Campanularia flexuosa*. The graphs in Figure 6.15 show changes in the growth rate of *C. Hexvosa* and in the permeability of its lysosomal membranes following an increase in the concentrations of these ions. The figures are expressed as a percentage of those in the controls.

Exercise 6.8 *continued*

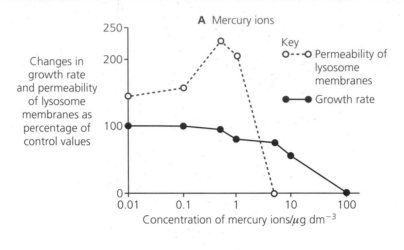

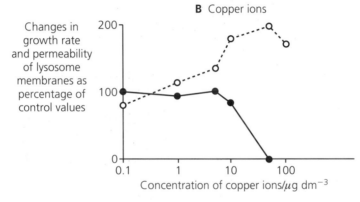

Figure 6.15 *The effect of the concentration of (A) mercury ions and (B) copper ions on the rate of growth and the permeability of lysosome membranes*

5   Suggest how the controls in this investigation should have been treated.

(1 mark)

6   (a)   It has been suggested that measuring the permeability of lysosome membranes may be more useful than measuring growth rate when monitoring copper pollution. Describe the evidence from graph **B** which supports this suggestion.

(2 marks)

(b)   Use Figure 6.15 to suggest one disadvantage of using changes in the permeability of lysosome membranes to monitor copper pollution.

(1 mark)